THE
FIRE CLUB

A dangerously humorous journey with
a regular crew of irregular firefighters.
Based on real events.

THE FIRE CLUB

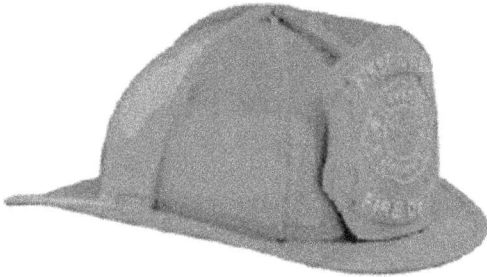

A dangerously humorous journey with
a regular crew of irregular firefighters.
Based on real events.

MICHAEL R. JASPERSON

ARPress
ILLUMINATING IDEAS
EMPOWERING VOICES

ARPress
45 Dan Road Suite 5
Canton MA 02021

Hotline: 1(888) 821-0229
Fax: 1(508) 545-7580

Ordering Information:
Quantity sales. Special discounts are available on quantity purchases by corporations, associations, and others. For details, contact the publisher at the address above.

Printed in the United States of America.
ISBN-13:   Paperback   979-8-89676-000-9
                eBook   979-8-89676-001-6

Library of Congress Control Number: 2024925152

This book is dedicated to my brothers and sisters in the fire service, who race into peril risking *their lives, their futures,* and *their dreams* to safeguard ours.

-Michael R. Jasperson

CONTENTS

CHAPTER 1

ANSWERING THE CALL

Tragedy stalked through the city of Brookfield. Sleek and agile, like a powerful cat, tragedy stole through the shadows, prowling the darkness, searching for the moment to reveal herself. Night drew tragedy closer to Brookfield's only firehouse.

Brookfield Station One lay dark and quiet. Inside, mirrored in the reflection of the polished brass bell were the sanguine fire engine and paramedic truck. Turnout coats, smudged with ash, hung ready on the handrails and open doors of the fire trucks. But the fire bell had not yet sounded the alarm.

Upstairs, the iridescent face of the dispatch radio cast its fractured reflection across the glossy, waxed floor providing the only light in the dormitory. Yellow turnout pants crumpled tightly around rubber fire boots sat beside each bunk. Five firefighters slept oblivious of their post.

3:14 AM.

The striker of the veteran alarm rose slowly as the electrical impulse warmed the well-worn solenoid. The clapper struck lightly against the thick brass bell emitting a soft, yet distinct *ding*. The firemen stirred. The warming striker drew back and hammered the bell with enthusiasm.

Rrrriiiinnnnggggggg!

The alarm bell's clamor reverberated hard against the red brick walls. From the first bunk, Captain Keith Hastings rose from his slumber. He rubbed the sleep from his tired face and slipped his feet into the rubber boots. He stood and raised the turnout pants to his waist and snapped the suspenders against his shoulders.

The voice of the male dispatcher boomed through the dorm. "Engine and Squad One. Residential fire alarm. Thirteen twenty-four Adobe Street. Cross Fiesta.Three-fourteen."

Captain Hastings ran his fingers through his wavy gray hair and scanned the dorm. Three of the firefighters were awake and in various stages of dress. Hastings briskly walked to the end of the row and kicked the rookie's bunk.

"Putnam, we've got a fire call!"

Startled, the young fireman sprang to his feet and hastily dressed. Captain Hastings called to the paramedics, Valdez and Razmunson, as they headed for the brass pole. "At this hour, think search and rescue."

The paramedics acknowledged with nods and slid down the pole.

The apparatus room came to life as the men leaped to their rigs and the fire dog commenced barking. Not needing acknowledgement from the captain, Engineer Barker flipped on the red lights and pointed the fire engine toward the far side of town.

As they crested the top of the mid-town knoll, the lights of Los Angeles flickered all around them like a sea of embers. Captain Hastings saw something that made him stiffen in his seat. In the distance, blotting out a section of spangling skyline, was a faint, flat cloud of smoke.

Captain Hastings knocked on the back window to Firefighter Putnam riding backwards in the jump seat. Hastings held his hand to his face and mouthed the word "mask". Putnam held a thumb up with his gloved hand and strapped on his breathing apparatus.

The fire engine's huge tires squealed around each corner as the engineer increased their speed. Turning through the entrance of the housing tract, Engine One was met by a thick bank of smoke.

The burning house was at the end of the long cul-de-sac. Orange flames glowed through the cracks of the tile roof, making the house look as if it were covered with briquettes.

"Attic fire," said the engineer.

"Tile roof," answered the captain.

A crowd of neighbors in various sleepwear lined the curb; some waving, some pointing, others spraying ineffective garden hoses into the windows of the burning house.

Captain Hastings keyed the radio mic and calmly spoke, "Stan. Keith…" He waited for the dispatcher to respond. The dispatcher eventually answered, sounding sleepy.

"This is a worker, Stan. Start a mutual-aid structure response from L.A. Fire," ordered Captain Hastings. The dispatcher acknowledged by repeating the order.

Turning his collar up against the bite of the cold night, Captain Hastings stood in front of the fire engine and surveyed the burning house like a warrior sizing-up his enemy. Smoke and fire blew horizontally from the attic vents, the fire had not yet vented itself through the clay tile roof.

Captain Hastings ordered Putnam to advance the hose line to the front door. Then, he turned to the crowd huddled on the curb. "How many people live in the house?" he demanded.

An old woman rushed forward, clutching the lapels of her bathrobe. "Robert and Maria are still inside the house," she cried. She looked back to her neighbors who nodded in agreement.

Without acknowledging the woman, Captain Hastings donned his air bottle and again studied the house, his face pulsating red with the flashing emergency lights. The attic glowed ardent orange but the roofline wasn't sagging. There was time for a search – a quick search. He watched Putnam advancing the attack line across the lawn and Valdez pulling a second, defensive hose line toward the adjacent house. On the front porch, Razmunson struggled with a body. The man's light blue pajamas gathered around his ankles, as Razmunson dragged his flaccid body to the front lawn. Black soot and saliva streamed from the man's nose and lips. His mouth gaped open, and his chest heaved with convulsions. His tongue was black with soot, his ashen face lies in sharp contrast to the green winter rye of the lawn.

Paramedics Valdez and Razmunson, encumbered with their turnout clothing and air bottles, checked the man for signs of life. Valdez forced the oxygen mask against the man's face, while Razmunson pressed his fingers to the man's carotid artery and searched for a pulse. Putnam, large and awkward, hovered over their shoulders. Captain Hastings pulled on Putnam's coat and pointed to the house.

Already breathing bottled air, Putnam hissed in mechanical gasps as he pulled the hose line to the front door. His breathing was rapid and deep. Both captain and fireman crouched on the porch stoop, taking a moment to study the house interior. Inverted mounds of smoke pressed down from the ceiling like the boiling underbelly of a storm cloud. Captain Hastings felt hot air blowing through the protective layers of his turnout coat. The limp fire hose stiffened with water pressure.

"Let's go!" shouted the captain, pointing into the dark menagerie of the smoke-charged house.

Inside the living room, the dense ceiling of smoke pressed further down. Nothing higher than the back of the couch could be seen through the inversion layer. The unseen yet palpable fire crackled and snarled overhead. As the firemen cautiously advanced, the searing heat bit into their backs. The captain pressed down on Putnam's shoulder, directing him to crawl.

Advancing the rigid fire hose was difficult from a crawl. Putnam jerked impatiently on the obstinate hose. From his prone position, Captain Hastings relayed more hose to feed Putnam's advance. Dragging the unyielding hose, their lifeline, the firefighters crept deeper into the house.

Passing by the living room couch, Captain Hastings groped the cushions hoping to find a body. His elbow broke the heat plane of the inversion layer, causing his skin to sting as it pressed tight against his turnout coat. He knew their time inside was limited. Periodically, the firemen shuffled on their hands and knees for expediency, but the superheated air above forced them back into a crawl.

Turning the corner at the china hutch, Captain Hastings pulled extra hose and relayed it to Putnam who was disappearing down the hallway. A large piece of ceiling plaster fell, sending lamps and other unseen items crashing to the floor. Pieces of flaming attic insulation rained down from the darkness above. The floor was speckled with glowing embers of smoldering insulation. Hundreds of miniature smoke columns rose to join the descending smoke ceiling.

"Cool the attic," shouted the captain.

Putnam stared blankly.

"Shoot up," Hastings shouted again, motioning up to the ceiling. Putnam jabbed the hose nozzle into the overhead abyss, causing his gloved hand to sting as it pierced the heat plane, and then pulled the handle. The water made a slapping bang as it disappeared into the smoke. Instantly, a fist of steam exploded over the firemen, singeing their ears and cheeks through their protective hoods. A slab of plaster fell directly on Putnam's head, offsetting his helmet and air mask. Air hissed as it escaped around his cheeks. fell directly on Putnam's head, offsetting his helmet and air mask. Air hissed as it escaped around his cheeks. He scrambled frantically to readjust his mask as hot, acidic smoke burned his throat. He retched uncontrollably. He forced his air mask hard against his face, sealing the edges and stopping further air loss. He paused for a moment until his nausea partially subsided.

Resuming their search, inching toward the end of the hall, past the bathroom and the two smaller bedrooms, they reached the master bedroom. Large sections of ceiling plaster fell randomly from above. Each direct hit drove the firemen hard against the floor. After the plaster fell, burning pieces of insulation rained like fiery brimstones from the inferno above. Glowing insulation dotted their route. Putnam shot water down the hall, darkening the embers.

Smoke obscured the flames overhead, but the heat was telltale that the fire ran the entire length of the attic. Putnam shot a burst of water into the blackness above. The fire answered back with explosive steam, knocking both firefighters hard against the floor.

The fire gnawed at the attic rafters and seemed to growl from all around them. They crawled quickly through the master bedroom doorway, dragging the hose line—their only direction out—with them. A short distance into the bedroom, the hose stopped.

"More hose," yelled Putnam, his words muffled as he screamed into his facemask, his voice showing signs of panic. The inside of his facemask dripped with condensation—he was hyperventilating.

"Slow your breathing. Save your air," ordered the captain. Hastings pulled on the hose line. It did not budge. He braced his feet against the doorjamb and arched his back, straining against the stubborn hose. Back in the living room, the fire hose tightened against the French legs of the china cabinet, buckling them, and toppling the hutch with a loud crash. For a moment, delicate tinkling continued as a cascade of fine crystals and porcelain figurines slid from the cantered shelves and smashed on the floor. The hose gave a few more precious feet.

Putnam advanced into the bedroom and felt the top of the kingsize bed. First, the pillows gave him a startle, but he found nothing.

Putnam's low air warning bell began to ring. He hastened his pace until he reached the far wall of the bedroom, but again, found no body. He returned to the captain. "Nothing," he shouted at Hastings.

"Let's get out of here," ordered the captain, pulling on the fireman's sleeve and pushing him back toward the hall. Putnam tugged at the nozzle, but the rigid fire hose was wedged between the bed and the back wall.

Putnam yanked on the stubborn hose, but it did not budge.

"Get out!" shouted Hastings, jerking his thumb toward the doorway. "Hurry!"

Putnam dropped the hose and scrambled down the hall, blindly groping his way, sliding his gloved hands along the hose line. He could scarcely see the hose at half an arm's length in the wavering beam of his flashlight. Even as a rookie, he knew losing contact with the hose and getting lost would have lethal consequences.

Something large crashed ahead of the firemen. Captain Hastings recognized the sound as falling roof tile. The roof trusses were collapsing. Crawling frantically, Captain Hastings pushed his helmet against his fireman, driving him forward. Putnam sprinted on his hands and knees, ignoring the bite of fire on his back.

By now, both of their low air warning alarms were ringing. Again, to the left, there was another crash. Then, to the right, another. The attic rafters, fatigued from the inferno, creaked and groaned under the tons of roof tile they supported.

A section of ceiling plaster fell onto Putnam. He floundered and winced as flaming insulation wedged between his air bottle and the nape of his neck. It was impossible to brush the embers away or cool them because he had left the hose nozzle back in the bedroom. Putnam stood to run but was driven to the floor by the intense heat. Instinctively, he slapped his hands to his stinging ears, rupturing the blisters.

Putnam again rose to his knees but was knocked down by a blow from behind. The captain had lunged forward to free himself from a pile of plaster and burning insulation. Regaining his balance, Putnam scrambled down the hallway, sliding his hands along the hose and praying it would guide him out of their hell.

Visibility was zero. Putnam banged his head into the wall as the hallway turned, then found the hose coiled into loops under the debris of the china cabinet. Although his skin felt as if it were on fire, he had to pause over the coil of fire hose and carefully trace its direction. The hose coil was too wide to reach across and shortcut the process, so he trapped the hose between his knees and followed it in two complete circles. Air was hard to inhale, a sign of a near empty bottle. The ringing low-air alarm fueled his anxiety.

They had two minutes of air left—tops—and the way Putnam was breathing, probably less. The beams from their flashlights, only inches long, illuminated nothing but the hose itself. Large piles of plaster and glowing insulation obscured the hose line. Putnam shook the hose free from the debris before scurrying into the darkness.

For a moment, Captain Hastings stopped. His low air alarm clanged desperately. He reached behind his back and held the bell with his gloved hand, muffling the ring into a frantic click. Then he heard it again; it was a snore, coming from his right. Hooking the hose with the toe of his boot, he stretched toward the sound coming from the darkness. He fumbled blindly along the wall, then pushed back the bathroom door. Faint in the dim beam of his flashlight was a woman's bare foot. He leaned back into the hall to call for help, but Putnam's alarm bell had faded into the darkness. Captain Hastings fumbled at the woman. He tugged on her chiffon nighty. It tore.

The unconscious woman weighed about two hundred pounds. She was too big to carry. Captain Hastings stood in the relative cool of the bathroom, straddled the woman and pulled her by the arms. Her snore changed pitch with each flaccid, flop of her head. He shuffled into the hallway, dragging the woman and burning his back with every move.

Desperate for leverage, he sat straddling the woman's head. He hooked his bulky leather gloves under her flabby armpits and dragged her down the hall. Pieces of smoldering insulation fell on her body, melting her nightgown, and sticking to her skin. At first, he brushed away the raining embers, but ceased at the futility.

Each time Hastings leaned forward to pull the woman, he crimped the inlet hose to his face piece, stopping his air supply. He had to time his breathing with each pull. The low air pressure made his breathing more difficult. The air-activated alarm bell tapped slower and softer, then fell silent. His next inhalation sucked the mask against his face. He was out of air. He fumbled at the hose connection on the regulator. His gloved fingers had no dexterity for the air tube's knurled fitting. His lungs ached for oxygen.

Yanking a glove off, his hand burned as he unscrewed the hose from the regulator. He fought to control his urge to breathe as he stuffed the open end of the inhalation hose into his turnout coat. He gasped a breath. Hot air burned the back of his throat; the sting of acid filled his mouth.

With his alarm bell silent, he could hear the angry roar of the fire. It echoed all around him. Ignoring the fire's bellow, Captain Hastings tugged again on the flabby arms of the corpulent woman. Sliding backwards, towing the woman, his head in the superheated air, the blisters on his ears burst filling his ear canals with cool liquid. He backed into the fallen china cabinet. Pressing against the cabinet, using the remaining strength in his legs, he pushed it aside with a crash. He tugged at the woman, pinching into her flabby arms, dragging her over the broken plates and ceramics.

His lungs burned. His head pounded. He dreaded each breath. Strands of mucus poured from his mask as he pulled it off to vomit.

Drunkenly, he replaced his facemask and pulled again on the woman's flaccid arms, but his fingers could no longer hold her slippery, rendering flesh. He collapsed to the floor and stared blankly at the broken porcelain figurines pressed against his facemask.

Captain Hastings paused for a moment to rest. *Just for a moment*, he told himself. As he lay crumpled and spent on the living room floor, the roof trusses gave way, and tragedy, riding the falling rafters, claimed her prey.

CHAPTER 2

CAPTAIN COLE WALKER

Cole Walker slowly opened the door to the fire department's automotive shop. The mechanics were not on duty, their work bays shadowed and silent. He slid inside and turned on the shop lights, illuminating a huge array of fire apparatus. He strolled amongst the fire engines and ladder trucks, soaking in their smoky redolence, and noting their scraped and blistered paint, scars of veteran service. He sat in a jump seat and held an axe, wondering what it was like to actually fight fire.

The closest Cole Walker had come to fighting fire was the mock fire simulations in the controlled environment of the fire academy. In the academy, he excelled in his studies and was the class valedictorian. On graduation day, the Fire Chief rewarded Walker's academic prowess by drafting him into the administrative offices where he had remained sequestered for seven years.

Cole mindlessly strolled among the apparatus, running his fingers along the smooth contours of a new pumper when he was startled by his pager's vibration. He gave his watch a quick check - *0800*. He bolted for the door and ran up the long drive until he came within view of the headquarters building where he slowed his pace to a standard business gait.

When Walker reached the entrance, he pulled the glass door open and stood aside, allowing two secretaries from the Personnel Office to enter first. The threesome exchanged pleasantries as they strolled down the long, stark hallway until the women turned into their office. Cole surveyed the empty hall and resumed his run. He ran down the corridor until he neared the first open office door, slid to a stop, walked nonchalantly past the doorway and casually waved to the people inside. Once out of their sight, he sprinted to the next door, slid to a stop, and repeated the scenario.

Cole's office was the last one at the far end of the dismal corridor. After his final dash, he slid on his patent leather shoes into the Public Information Office. Pluggy, a red plastic fire hydrant, blocked the doorway. The hydrant turned and rolled toward Cole, forcing him to step back. Two big eyes opened from its fluted bonnet.

"You're late," scolded the hydrant in a digitized voice. "Now that you're a captain, you're supposed to set a good example for the rest of the crew." The hydrant blinked its plastic eyes and rolled backward into the office.

Inside the PIO stood Red Hall, Cole's best friend and fellow headquarters detainee. Red Hall was also in his late twenties and had the same thin physique as Cole, but had dark, wavy red hair. Like Cole, Red Hall, had been transferred into headquarters directly from the academy seven years earlier. Over the years, Cole had been shuffled between several different offices. Red had remained in the stockroom.

Red Hall held the remote-control joystick and had the virtual reality helmet strapped on his head. He stood disoriented, facing the corner, still looking through the eyes of the fire hydrant. "I paged you because I figured you were in Automotive daydreaming again."

"I wasn't daydreaming," defended Cole.

"It sure looks like daydreaming."

"I was *not* daydreaming! I was preparing for field duty through equipment familiarization."

"Why?" asked Red. "Why do you do that to yourself? For seven years, you've been saying that. And for seven years I've been telling you, 'You ain't going nowhere.' It's about time you get a life."

"This *is* my life!" shouted Cole. He composed himself, then spoke just above a whisper. "I am a firefighter, but I've never fought a fire. I have never ridden on a fire engine. I'm rotting behind this desk."

"Cole, you need a hobby, or a girl, or some sort of diversion."

"I want to fight fire," said Cole, pounding his fist on his desk. "And I will do nothing else until I get an engine assignment."

"Cole, you know neither of us is ever leaving headquarters. The closest you'll ever get to fire is blowing out the candles on your retirement cake." Red saw the disappointment on Cole's face. He offered a consolation. "Look, Cole, if you figure a way to tunnel out of this place, leave the hole open. I'll crawl out with you."

Cole knelt at the hydrant and grasped it between his hands. "When did Pluggy get here?"

"Just this morning," said Red Hall, struggling to pull the monitor off his face. He fumbled blindly at the straps until he lost his patience and forcibly pried the monitor off his head. Red plucked the headset from his smashed red hair and held it under his arm like a football helmet. The nylon straps left red depressions on his pale forehead.

"Look here. This is great," said Red. "It talks through here," he said, tapping on the hydrant's black mouth screen. "And the nose is a remote camera. When you move the lens, the eyes follow." Red rotated the joystick, rolling the hydrant's eyes from side to side.

"I know. I ordered it," said Cole with disinterest.

"The chicks dig it," said Red.

Cole rolled his eyes, accenting his disgust.

Red Hall ignored Cole's lack of enthusiasm and continued his spiel. "Well, Mr. Know-it-all, did you know when you maneuver Pluggy next to a woman in a short skirt, you can look up her dress?" Red's freckled face beamed, obviously proud of his discovery.

"Gimme that," said Cole, snatching the control unit from Red's arm. "I ordered Pluggy for the School Education Unit... not as your sex toy. Besides, if we're ever going to get out of the admin building, we can't get caught doing stupid stuff like that."

Cole flopped into his desk chair and turned his attention to the remote-control unit. He then strapped on the video headset and spoke into the flex microphone. His voice came out of the hydrant sounding like a ten-year-old with his nose pinched. Cole pushed the joystick forward, crashing the hydrant into the wall. He gave the joystick a few practice moves, then sent Pluggy whirling down the hall toward Chief Pierce's office.

The Chief's secretary, Sandra Goodale, was at her desk, trying to decipher the Chief's scrawl from a crumpled piece of paper before she transcribed it into a formal letter. She looked up from her work when she heard, from down the hall, a woman gasp and several books fall to the floor. Shortly afterwards, a fire hydrant lurched through the office door and slammed into her desk. The hydrant backed up and came to the side of her chair.

"Good morning, Sandra."

Nobody called her 'Sandra' except Cole Walker. Sandra Goodale was nearly twice Cole's age and he treated her with as much regard as his own mother. She spun her shapely legs from under the desk and leaned to the hydrant, softly caressing the top bonnet. "Is that you in there, Cole?"

"How did you know?" replied the hydrant, obviously disappointed. Positioned in front of her knees, Cole diverted the hydrant's eyes from looking directly up her skirt. "Has the Chief processed the transfer requests yet?"

"No, Cole. I'll send you the first copy as soon as I get it, just like I did last month and every month before." Sandra then consoled Cole. "I put your transfer request on the top of the stack, but Chief

Pierce has been busy with the Brookfield annexation proposal. He's spent the last couple days down at Brookfield City Hall. In fact," she paused to check her phone; line one was still lit, "he's been talking to Mayor Jenkins for over an hour."

She patted the hydrant sympathetically. "Don't worry, Cole. Something will work out."

"Thanks, Sandra," said Pluggy's digital voice. Then the eyes of the little hydrant turned down toward the floor, quickly past her knees. Within the periphery of the camera's field of view, Cole saw the tips of shiny black shoes. The camera lens ran up the sharply creased uniform pants until suddenly, Chief Pierce's face filled the monitor. Cole jumped in his PIO office.

Chief Pierce rapped his knuckles between Pluggy's eyes. "Walker, is that you?"

"Yes, sir," the sheepish voice replied.

"Come to my office," said the Chief. He put his thick finger on Pluggy's nose and filled the TV monitor strapped to Cole's face. He pushed the hydrant backwards across the office and into the hall. "No toys!" Pierce shouted and slammed his office door.

Back in the PIO, Cole shot to attention and jammed his thighs under his desk. He wrestled with the virtual headset but was unable to release its grip on his face. Red slapped Cole's fumbling fingers and unsnapped the buckle.

"The Chief wants to see me," said Cole, as he threw the headset to Red and darted out the door.

Red Hall splayed his bandaged fingers, "Tell the Old Man to transfer me out of the stockroom before I bleed to death from paper cuts."

Cole brushed his fingers through his matted blond hair and scrambled down the hall. The chief's office was at the opposite end of the long hallway. Recessed, fluorescent tube lights gave the corridor a ribbed effect like the inside of a large serpent; —a serpent that had swallowed Cole fresh from the academy. At the other end of the corridor, passing in and out of the light, marched a man. Though not fully illuminated, the militaristic stride and the tightly cropped hair was the insignia of only one man: Captain Decker.

Captain Decker, a former Marine drill instructor, had been Cole's training captain in the fire academy. His starch-hard demeanor and harsh discipline earned him the 'most hated instructor' status. After Walker's academy graduation, Chief Pierce transferred Decker out of Training because he was too hard on the recruits. Rumor had it he had been released from the Marine Corps for the same reason. Captain Decker loathed Cole Walker.

Cole wasn't exactly sure why Decker hated him. It may have been because he was the youngest recruit in the academy or because he publicly corrected Decker on the flashpoint of gasoline, or because he once opened the wrong pump discharge gate blowing Decker into the rancid water of the drafting pit. Whatever the reason, Decker's anger soured with time, leaving nothing but venom for Cole.

Though it had been seven years since his old training captain had barked orders at him, Cole could still feel Decker's sour, dank breath blasting against his face. They met midway in the corridor. Neither man offered a handshake.

"Walker," sneered Captain Decker. "I hear it's now, *Captain* Walker. I should congratulate you. Getting promoted is a tough thing to accomplish from behind a desk.

Cole made no comment.

"It looks like putting you in headquarters was the right move," smirked Captain Decker.

"How's that?" asked Cole.

"It was my idea to send you to H.Q. I told the Old Man that paper shuffling was more your style. And now you've promoted to Captain. Looks like I was right."

Captain Decker stepped around Cole who stood rigid, his fists clenched, burning with anger at Decker's revelation. For seven years, Cole had assumed Chief Pierce had put him in the administrative offices. Now, it all made sense. The Fire Chief was too removed to single out any particular recruit. The past seven years of drudgery was Decker's doing.

Cole trudged furiously down the hall. He could not shake his anger as his thoughts boiled with Captain Decker. He met Pluggy coming back the opposite way.

"Good luck," Red's playful digitized voice from the fire hydrant now blocking Cole's path. Cole stepped right. Pluggy went left, blocking his way. Cole stepped left at the same instant Pluggy swerved right. Perfectly choreographed, they blocked each other's way with simultaneous sidesteps. When they got close enough for contact, Cole shoved the little hydrant, sending it skidding across the slick tile floor.

Red Hall, dizzied by the spinning monitor, fell in the PIO office. "Hey! What was that for?" whined the hydrant, rocking on its back. Cole stepped over the convulsing hydrant and continued down the hall. He paused outside the double oak doors, took a deep breath, straightened his posture and entered the Fire Chief's office.

Sandra Goodale was sitting at her desk concentrating on the Chief's scribbled note. She didn't notice Cole enter. Cole cleared his throat. When Sandra looked up, the corners of her stern lips curled into a grin. She greeted Cole and picked up the phone in one fluid motion. She buzzed Chief Pierce and acknowledged his response. "He's waiting for you, Cole. Good luck."

Before entering the Chief's office, Cole cleared his throat and brushed back his straw-like hair, exposing the reddened strap marks across his forehead. He gave Sandra a thumbs up then pulled open the heavy oak door.

Chief Pierce was standing at the office wall, facing a large map of the southland. The center of the map was Los Angeles. The Chief spoke without turning around. "Captain Walker, tell me what you see here."

The ambiguous question caught Cole off-guard. "I'm sorry, Chief, I-I don't understand."

"Right here," repeated the Chief, louder, circling the entire map with his finger. "What does this look like to you?"

Still confused, Cole shrugged his shoulders. The Chief's stocky physique was intimidating, and his vague query could mean anything. Cole stammered, "I'm not sure I understand th…"

"A donut," Chief Pierce interrupted.

"A donut?" "Yes, Captain Walker. A donut," snapped the Fire Chief, turning to Cole with an angry glare. The Chief framed a smaller section of the map with his hands. "What is this, Captain Walker?"

"Brookfield?" offered Cole.

"That is not Brookfield, Captain Walker. It's a big, goddamn hole right in the middle of my fire department." The Chief's voice increased in volume as he turned again to face the map. "The Los Angeles Fire Department's consolidated fire protection district consists of 60 incorporated cities." Chief Pierce began tapping random sections of the map, naming the cities contracting fire protection from his larger fire department. "Rosemead, Altadena, Malibu, Lakewood, Bell, Cerritos, Artesia…" His finger made a loud *pop* each time he struck the map. "…Walnut, El Monte, El Segundo, Westwood, Lynwood…" Chief Pierce turned and stared hard at Cole. "But not Brookfield."

The Chief pointed his thick, sausage-sized finger to the top of the map. "What does that look like to you, Captain Walker?"

"Kern County?" answered Cole, not sure where the conversation was going.

"I know *what* it is, Captain Walker. I asked, 'What does it *look* like?' Chief Pierce made no attempt to hide his irritation.

"A bear?" offered Cole, squinting his eyes and grimacing at his idiotic answer.

"Right!" exclaimed the Chief, now pointing his sausage finger at Cole. "And what does it look like with a bear standing on top of a donut, Captain Walker?"

"Well, I guess…"

"A circus act, Captain Walker. It looks like a goddamn circus act. Chief Daniel's fire department looks like a bear and mine looks like a donut. Do you have any idea how embarrassing that is?"

Cole made no attempt to answer, deciding that he wasn't supposed to. Chief Pierce turned to his desk and held up every example he could find of the Los Angeles Fire Department emblem—paper weights, survey folders, business cards, letterhead and manuals—each outlining the Department's fire protection district into what Cole could only recognize as a donut.

From the stack of examples piled on his desk, Chief Pierce pulled a folder labeled *Brookfield Annexation Proposal*. The proposal was Cole's most recent project in PIO. Cole had a hard time taking his eyes off the County emblem that now looked amazingly like a donut. His fixation was broken when Chief Pierce slapped the annexation folder on top of the pile.

"Brookfield accepted our proposal for fire protection," said the Chief. His voice was back to its normal range, but Cole detected a hint of concern.

"That's great, Chief," said Cole, smiling and trying to lighten the moment. "Looks like you got your donut hole," he said, grinning, proud of his analogy.

The Chief's face scowled. He sat in his leather seat and motioned for Cole to sit in the wooden chair opposite his desk. The room was silent. The leather chair creaked as Chief Pierce leaned forward, resting his thick forearms on his large oak desk. "Unfortunately, Captain Walker, there are some limitations to the contract. Mayor Jerkins…"

"*Jenkins*, sir. His name's pronounced Jen-kins." Cole's voice trailed off under the weight of the Chief's stare. Obviously, the Chief did not appreciate the interruption, or the correction.

Chief Pierce continued, holding his stare on Cole and clearing his throat for emphasis. "Mayor *Jerk-ins* is not in favor of our fire protection. Luckily, he was outvoted by the city council, three to two. Unfortunately, instead of signing a ten-year contract, he was able to convince his fine councilors to accept our offer on a one-year trial basis with an escape clause. Should our services not meet their expectations, they have the option of returning to their pathetic, inept, little fire department, leaving me with a four-point-six square mile hole in the middle of mine!"

Cole knew the importance of annexing Brookfield. After all, he'd written the proposal. Filling the hole in the center of their jurisdiction would not only improve the look of the Department's emblem, but the surrounding contract cities would receive quicker fire response with Los Angeles Fire Department engines and truck companies neighboring on all borders. The Brookfield annexation was like putting in the last piece of a giant puzzle.

"What's your plan, Chief?" asked Cole, sitting on the edge of his chair.

"Last month, Brookfield lost one of their captains in a house fire," said Chief Pierce.

"Captain Hastings," interjected Cole. "I attended the funeral and sent flowers on your behalf."

"Right," continued the Chief, half-embarrassed, half-annoyed. "The Brookfield town council wants their fire personnel left intact should this annexation fail. Although, they will allow one Los Angeles fire captain to fill the existing vacancy."

The gears in Cole's mind spun wild. Filling the captain vacancy in Brookfield would create another vacancy somewhere else in the Department, a hole that perhaps he could fill.

"Have you decided who to send?" blurted Cole, unable to contain his excitement. "I know Captain Decker would want the spot. He has tons of experience." Although Cole didn't owe Captain Decker any favors, transferring him to Brookfield would open up a coveted captain's position on Ladder Truck 45— a spot that perhaps Cole could fill.

Chief Pierce leaned back in his chair and put his hands behind his thick neck. "I know. Decker submitted a transfer request this morning just in case Brookfield accepted our offer. And, yes, he is one of my most experienced captains, but he's…" The Chief's statement trailed off as his eyes searched the ceiling for the right word.

Cole couldn't contain his enthusiasm and tried to complete the Chief's sentence. "Too intense... Abrasive... Forward... One-way... Egotistical... Militaristic... Rough..."

"Ass!" proclaimed the Chief, slapping his thick hand on his desk. "Captain Decker is an ass! He'll have Mayor Jerk-ins in a headlock at their first introduction." He held his eyes on Cole. "I was thinking of someone with more political savvy."

The suspense was too much for Cole. "Who?" he probed, reeling with curiosity. He repeated the question, playfully demanding, "Who... Who?"

"You."

"Me!" Cole shouted, pressing back into his chair. He checked the Chief's eyes, hoping they would reveal his joke.

"No... not me, Chief," Cole said, shaking his head. His pale pallor changed to cadaverous. As much as he wanted a field transfer, as much as he wanted out of headquarters, Brookfield was too large a leap. It was too much of a career-ending, political hotbed.

"I've only been a captain for three days," bleated Cole. "I've never worked in the field!"

"Don't worry, Walker. You're a quick learner. Besides, you're a smooth talker." Chief Pierce leaned forward across his desk, staring down at Cole who pressed back against his chair. "I need someone to appease Mayor Jerk- ins. After all, who knows Brookfield better than you? You wrote the book."

Cole's retort reeked of sarcasm. "Let me recap, Chief. You want me, a captain of three-days' experience to take the place of a dead, veteran captain in a city that has only lukewarm interest in our fire protection?"

"Precisely," said the Chief, smiling at Cole's grasp of the situation.

"The crew will love me," mumbled Cole, continuing his sarcasm. "When does the transition take place?"

"Next C-shift."

"That's this Monday!" Cole said, almost shouting. "That's in three days! And what if the deal falls through? What if I can't convince Mayor Jenkins to extend the contract?"

Chief Pierce slowly raised himself from his chair, crossed the side of his desk and stood over Cole. He leaned close to the young captain before speaking. "Failure is a subject we don't want to discuss." Chief Pierce forced a broad smile and extended his beefy hand. "Congratulations, Captain Walker."

In one motion, Chief Pierce shook Cole's hand and pulled him out of his chair. He put his thick arm around Cole's shoulders and steered the stupefied young captain out of the office and down the hall. "It's time for the staff meeting," said the Chief, and flashed Cole a smile.

When the two men entered the conference room, they found it crowded with headquarters personnel all wearing party hats. "Surprise!" The room exploded with cheers. The large conference room was decorated with colorful streamers and balloons. The ladies from Procurement had decorated a cake with a plastic fire engine tucked halfway inside a graham cracker firehouse. A white banner spanning the room read, *Congratulations Captain Walker.*

The room was packed with everybody from headquarters who all took turns congratulating Cole on his promotion. Cole, still numb from his meeting with the fire chief, feebly shook hands with the blur of people. Cole shuffled through the well-wishers to Pluggy who was sided next to a tall secretary in a short skirt. Cole slapped the top of the hydrant and pushed it into the corner. "Red, we've got to talk. Let's get out of here." Cole spun the little hydrant around and pushed it toward the exit.

Just before reaching the doorway, they ran into Chief Pierce. The Chief was holding a piece of cake in one hand and a fork in the other, blocking their egress. He pointed the fork at Cole, "Now that you're leaving PIO, do you have a recommendation for your replacement?" asked the Chief, turning his attention back to his slice of cake.

Cole heard Pluggy's casters spin and the whirl of the camera lens as it focused on him. Pluggy's eyelids batted up and down, clicking and winking. The little hydrant jockeyed back and forth, bumping into Cole's leg.

"Red Hall," Cole replied.

"Hall... Hall..." said Chief Pierce, trying to place a face with the name. "Isn't he the freckle-face, red-headed kid in the stockroom?"

"Yes, sir."

"Why does he have all those bandages on his fingers?" asked the Chief. Pierce frowned hard and allowed his stare to bore into Cole. "Are you sure he's the right man for the job?"

"Yes, sir, as much as I'm the right man for Brookfield.

CHAPTER 3

FIRST IMPRESSIONS

Thirty minutes passed, while Captain Cole Walker sat in his car across the street from the Brookfield fire station. His anxiety was fueled by the most perplexing gamut of emotions he had ever experienced. He was happy to be assigned to a real fire station, yet deathly afraid to assume command. Nervous and anxious added up to nausea.

The provincial firehouse was a picturesque, two story, red brick building framed with white masonry corbels and ornate cornices. The old swinging doors had been upgraded with modern sectional roll-ups. From the bright white flagpole with the brass eagle on top, to the old oak front door, the firehouse was more beautiful than any other he'd seen.

A large granite plaque centered in the peak above the two bay doors was inscribed with fancy English scroll:

BFD #1

Est. 1909

Leaning against the side of the station was a flimsy plastic sign. Cole knew what the sign said without reading it: *Los Angeles Fire Department – Station 61*. He had seen several others like it get slapped up after annexations. It was a shame to deface such a beautiful building with the Department's cheap replacement for quality craftsmanship. His stomach twisted at the irony.

Cole fidgeted nervously with the red plastic fire engine dangling from his ignition key. Again, he tried to gather the courage to drive in, but worried how the firemen would receive him. He expected them to be uncomfortable with a new captain. So, he planned to give them loose reins until they acclimated to him. He turned to the boxes of training manuals on his front seat. *Too late now*, he thought. If he wasn't prepared by now, it was too late to cram. Bolstering his courage, Captain Cole Walker started his car and turned into the long narrow driveway leading to the rear of the firehouse.

Unlike the front of the station, the back was less ornate. The rear yard was shadowed by towering, old, windowless warehouses abutting the property line, making the back lot extremely private. The doublewide bay door was open, but the firemen inside were too busy to notice Cole drive up.

The first to hear the car door shut was the station mascot, who rushed to Cole with a low growl. The dog, a pit bull-terrier mix, pressed Cole against the car until he was satisfied with Cole's scent. Conveying his approval, the spotted dog sat and wagged his stump of a tail.

The next person to notice Cole was the fireman who had been polishing the brass pole. He smiled instantly, set down his dripping can of polish and hurried across the apparatus room. The fireman was large, near three hundred pounds, and built like a boxcar. His hands were caked white with brass polish; his uniform pants were speckled with drips of the same.

"Hello. You must be Captain Walker," said the fireman, clasping Cole's hand between his two thick hands, transferring polish to both sides of Cole's hand. The fireman shook Cole's hand so vigorously that his vision blurred. "I'm Terry. Terry Putnam, the Boot."

Cole noticed the pinkish-red mottled tips of Putnam's ears — obviously healing blisters. With formal pomp, Putnam introduced Cole to the dog. "Captain Walker, meet Bleve. Blev, this is Captain Walker."

"Call me Cole," Cole said to both.

"Let me take those," said Putnam, as he pulled the bright yellow turnouts from Cole's arm. "I'll put them by the engine. Wanna meet the guys?" asked Putnam as he waved for Cole to follow him across the apparatus room.

The station was larger than most firehouses of its vintage. The apparatus room was deep and could easily hold four fire engines. Tall walls of dark red brick rose to the heavy timber beams spanning the ceiling. The room smelled of polish and cleaners with the familiar scent of smoke. Cole couldn't resist turning full circle to absorb the station's ambiance. Putnam stood patiently at the kitchen door, waiting for his new Captain to stop spinning. Two firemen were inside the kitchen, one seated at the table, one leaning against the counter. "These are our two paramedics, Jerry Razmunson and Richard Valdez," said Putnam. The two men offered nods as salutations.

Cole crossed the room to the first paramedic who had short, untamed hair jutting out in all directions. He held his hand to Paramedic Razmunson who remained seated, "Call me Cole."

"Thanks," replied Razmunson, unenthused.

Next, Cole walked to Valdez who made no attempt to meet him halfway. Valdez was the shortest of the firemen, only as high as Cole's shoulders. His protruding forehead, jutting lips and pencil-thin mustache gave the little fireman a suave, Neanderthal quality.

"Pleased to meet you," lied Cole.

"Likewise," Valdez lied in return

"Want to see your office?" interjected Putnam, knowing he wasn't breaking up their conversation.

"Sure," said Cole. "See you later," he said with no response from the paramedics.

"Up front is the captain's office," said Putnam, as they walked to the front of the apparatus room. Putnam pulled open a heavy oak door and waved for Cole to enter first.

The office was a tiny room with barely enough space for a desk, file cabinet and small bookshelf. The only view outside was through the large window in the front door overlooking Main Street. The walls were decorated with a collage of black and white photographs of fires and other disasters. A horrendous vehicle accident hung next to a burning factory, next to the charred remains of a house, next to the picture of Mayor Jenkins. The largest picture occupied the premium position directly in front of the desk. The portrait was of Captain Keith Hastings, the same one Cole had seen in front of the casket at the fallen captain's funeral. It still had a black sash draped across the top.

"Captain Hastings," said Cole, acknowledging the picture. "I'm sorry. He must have been a wonderful man."

"The *best*," said Putnam, holding his trance on the picture. "He was the best."

The enveloping silence was crushing. Cole turned around in the small office, making obvious inspection of the contents. He ran his finger along the books and twirled the handle of the pencil sharpener.

"Wanna meet your engineer?" asked Putnam, shaking him from his trance and the awkward lull.

"Love to," said Cole, really meaning it.

Stepping again into the apparatus room, Cole studied the fire engine. It was a 1963 Crown triple-combination pumper, the same type Cole had trained on in the academy. Fire engines of that vintage had long since been retired as frontline rigs by the Los Angeles Fire Department, replaced with newer, more sophisticated apparatus.

The engineer was underneath the engine, only the tips of his boots showed from under the running boards.

"Rus, Captain Walker's here," announced Putnam.

"Uh-huh," said the engineer.

"I like your engine, Rus. Barker, right? It's in good shape," said Cole.

"She," corrected Barker. "*She*'s in good shape. Like ships, fire engines are *she*'s."

"Right, right," Cole said, speaking to Barker's boots. "I'll keep that in mind."

"I bet you wanna see your locker," said Putnam, still smiling.

"That sounds excellent. See you later, Rus," said Cole to the engineer's boots.

"Uh-huh."

Putnam led Cole upstairs, down the narrow hallway, past the dorm, past the shower room, and into the locker room. The room was small with narrow corridors between the narrow lockers.

"It's kinda tight in here," said Putnam, pointing out the obvious."Each shift has a row of lockers. Mine's next to Val's," he said, pointing to his right.

"Wow, what's all this?" asked Cole, stepping to the open locker. The inside of the locker door was covered with photographs, *Thank you* cards, garter belts, and other assorted memorabilia.

"This is Valdez's locker," said Putnam. "He posts items from his latest exploits for public display. He never closes his locker door." Putnam glanced over his shoulder and lowered his voice. "We call it the *Me-Shrine*," he said, nodding at the locker door for Cole to make a closer inspection. "Every photo has Val in it. Every item is one of conquest… garter belts, panty hose, panties…"

The collage included photos of Valdez sitting on a motorcycle with a beautiful woman; another photo was of Valdez on the beach with a different bikini-clad female, and another of Valdez on a jet ski, Valdez on snow skis, in a spa with twins, on a mountain peak.

"This doesn't fit," said Cole, pointing to the Girl Scout calendar.

"That's his prize possession," corrected Putnam.

Cole took a closer look. Scrawled in black felt pen across the canoe of girl scouts, *Richard, GREAT MEAL, Governor Atchison.*

Putnam told Cole about the time when the governor visited the station during one of his campaign tours.

"Governor Atchison?" interrupted Cole. "Now, Senator Atchison?"

"The same," said Putnam. "This was before his Senate bid. Apparently, when the governor visited the fire station, he stayed for dinner. It was a great photo op, connecting himself with the fire department and all. Valdez was the scheduled cook that day. The Governor ate with the crew. When the meal was over and the reporters started to thin, Governor Atchison hurried off. As he bustled out, Valdez grabbed the closest thing at hand which was the Girl Scout calendar and asked the Governor to sign it for him." Putnam whispered, "It's the center of the *Me-Shrine*."

Putnam then directed Cole to the first locker in the row. "This one is yours," he said, tapping his knuckles on the locker with *Hastings* labeled on the door.

The sight of the nametag made Cole swallow hard. He wondered if Putnam heard his audible gulp. He then thanked Putnam for the tour. The big man spun on his heels and trotted down the stairs with the dog following close behind. When he was out of sight, Cole opened the locker door. The locker was empty with nothing but a few wire hangers on the crossbar. He was thankful for that. Cole closed the door and looked over his shoulder to assure himself he was alone, then worked his fingers under the name plaque and tore it off, making his first move to take the place of Captain Hastings.

The dedication ceremony was scheduled for one o'clock that afternoon. The Brookfield Ladies Fire Auxiliary arrived early to decorate the station. With help from the firemen, the front of the firehouse was trimmed with red, white and blue ribbon that draped over the cornices and across the apparatus doors. Hundreds of feet of colorful streamers on red nylon cord cascaded down the station face, flowing inside the apparatus room, along the walls, and across the front of the pastry-laden card tables. Patriotic ribbons were meticulously tied across the center of each refreshment stand, cleverly hiding the over-stuffed, cellulose legs of the Ladies Auxiliary members as they sat wedged underneath.

The transition of fire protection was the biggest event to hit Brookfield since Governor Atchison's visit a decade earlier. All of the Los Angeles news media had been invited to the dedication ceremony, but only the reporter from the *Daily Brookfield* and Jennifer Preene from *TV-3* were in attendance. They began their coverage by quoting PIO Officer Red Hall when he described the annexation as a "smooth move."

Fire Chief Allen Pierce was the first speaker of the day. Standing tall and bullish in his dress uniform, square chin and tight haircut, he was an excellent fire department personification. He spoke of pride and dedication and public service. He listed the equipment now available to the Brookfield citizens: Helicopters, Air Squads, Urban Search and Rescue Teams, Hazardous Material Units, Ladder Trucks, Foam Units, Compressed Air Units and a fire boat, which led to a chuckle from the audience. He then directed his speech to Captain Cole Walker, stating that he was chosen from a vast personnel pool of highly qualified individuals. The Chief added that Captain Walker was selected because he was the most knowledgeable, aggressive and respected captain in the Los Angeles Fire Department. Under the direction of Captain Walker, the citizens of Brookfield would receive the finest fire protection available. Chief Pierce then welcomed Brookfield as the sixty-first member of the Los Angeles Fire Department, wooing the audience into a standing ovation.

Following Chief Pierce, Mayor Jenkins spoke. His gaunt and ashen physique was in stark contrast to the fire chief. His skin was maculated with liver spots and his tightly trimmed mustache was pure white, except for the nicotine streaks trailing from each nostril. His tweed sport coat reeked of stale cigarettes, his shoulders anointed with flakes of dandruff. The mayor cleared his throat with a smoker's hack and commenced his platitudes and, true to his political archetype, rambled on and on about yielding to the pressures of economics. At the conclusion of his marathon preamble Mayor Jenkins introduced the firefighters. The audience clapped after each name, obviously annoying the mayor. He continued, "I'm sure most of you recognize these dedicated men as they have been protecting our fine city for many years and will continue to serve under the new logo for the Los Angeles Fire Department." He turned to face the firefighters, though held his stonecold glare on Captain Walker, "Gentlemen, the safety of Brookfield is in your hands."

The focus of the ceremony then shifted to converting *Brookfield Engine-1* into *Los Angeles Fire Engine-61*. The alteration was inexpensive at best, merely placing a magnetic sign, *E 61* over the handpainted decorative door insignia. Chief Pierce symbolically held the magnetic sign over his head like a boxer holding his prize belt.

The magnetic sign was considerably smaller than the fancy hand-painted gold-leaf artwork on the fire engine, so when the Chief pressed the square *E- 61* magnet over the engine's *Brookfield Fire Department Engine One*, the ornate scrollwork showed from around the edges of the magnetic sign like the legs of a smashed bug. Seemingly oblivious to its tacky appearance, Chief Pierce stepped back and admired his work as he led the audience in another round of applause.

Next, Chief Pierce had the distinction of inducting Brookfield as the sixty-first member of the consolidated fire protection district. He picked up the microphone, which was connected to the station's fire radio. Waving the microphone aloofly, as if it were a magic wand, he fumbled for the transmit switch. *Click.* "L.A. Dispatch. Fire Chief Pierce," said the Chief. He liked hearing his name over the radio.

The voice of a female dispatcher boomed over the loudspeaker."Chief Pierce, L.A. Dispatch. Go ahead with your message."

"L.A., Chief Pierce," he said, pausing to smile and nod at the crowd, "Change Engine Sixty-one's status from *out of service* to *available for assignment*. Also, send a test alarm to Station Sixty-one."

The dispatcher repeated the order. Chief Pierce faced the alarm bell. Following his lead, the mayor, the ladies of the Brookfield Ladies Fire Auxiliary, the small clique of media and assorted guests turned with anticipation to the antique bell hanging on the wall. A flash arced across the crowd as the photographer for the *Daily Brookfield* took a picture of the bell.

Ding…

Eager to applaud, yet confused by the lone bell tap, a few people offered a few scattered claps. Chief Pierce looked at Captain Walker, then at the bell, then back at Captain Walker. Cole shrugged his shoulders.

Ddddrrrriiiinnnngggg!

The dispatcher's voice boomed across the brick walls. "L.A. Dispatch testing with Fire Station Sixty-one. Welcome aboard, Brookfield."

The room erupted in applause and cheers at Station 61's first alarm.

As a final gesture, Mayor Jenkins and Fire Chief Pierce posed like allies with oversized scissors before cutting the ceremonial ribbon. The cutting of the single red ribbon brought to a close 86 years of inhouse fire protection, and Mayor Jenkins' control over the small-town fire department was officially severed.

After the last umpteenth picture for the *Daily Brookfield,* Chief Pierce and Mayor Jenkins parted company. The Chief latched onto the *Daily Brookfield* reporter and Mayor Jenkins stepped out front for a smoke. On his way through the station, he grabbed Cole by the elbow and pulled him close. "I'll be watching you, boy. The first time you screw up, I'll jump on your neck with both feet."

"There won't be any problems, sir," said Cole, ignoring the mayor's hostility. "In fact, I think you'll be pleasantly surprised."

"Not likely," said the mayor before giving Cole's elbow a shove and storming away.

Numb from his encounter with Mayor Jenkins, Cole headed to the refreshment tables for something to quench his dry throat. He stood in line behind a sheriff with an expanding midriff as he concentrated on balancing pastries on his overloaded paper plate. He turned suddenly, nearly bumping into Cole.

"Captain Walker, right?" said the rotund officer with a jolly voice. Cole could see a white tee shirt bulging between each straining button of the officer's taunt shirt. The sheriff balanced his mounded plate on his left hand, licked the crumbs from his fingers of his right hand, and then offered Cole a handshake.

"Parker. Stanly Parker," the deputy said, warmly. "Welcome to Brookfield."

Cole shook Parker's hand; his fingers were warm and moist. Officer Parker was the last remaining remnant of the Brookfield Police Department. He had been the police and fire dispatcher, but when Brookfield contracted out for law enforcement with the County Sheriffs, his dispatching position was terminated. After 23 years in an isolated dispatch cubicle, Officer Parker was dumped into field duty. Officer Parker politely excused himself and escorted his plate to a table.

Cole stepped up to the refreshment tables and scanned the array of pastries. All of the card tables were covered with cookies, cakes, pies and sweet breads. But, a glass of water was all Cole was searching for when he happened onto the ancient members of the Brookfield Ladies Fire Auxiliary.

The Brookfield Ladies Fire Auxiliary was a staple at all fire department functions. Dorothy Greene, Lilly Hoffman and Helga Broadenhoff were the auxiliary's charter members and sat wedged behind the tables before him.

Dorothy Greene was the first to stand, holding out her wrinkled hand like a dead fish. Cole politely shook her icy cold, liver-spotted hand. He flinched, and then blushed.

"Hello, Captain Walker. I'm Dorothy, President of the Brookfield Ladies Fire Auxiliary. We're still keeping our *Brookfield* name because we couldn't possibly service all of the handsome firemen in the entire Los Angeles Fire Department." She turned to the other two elderly ladies, who were cackling and nudging each other under the table.

Cole couldn't help but notice Dorothy Greene's protruding teethand gums. She reminded him of a baying mare as she whinnied with the other auxiliary ladies. When Dorothy Greene regained her composure and turned back to Cole, his eyes remained on her teeth.

"Captain Walker," said Dorothy Greene. "Captain Walker," she repeated, startling Cole from his fixation on her piranha-like overbite. He jumped back from his stupor; his cheeks flushed red again.

"Please call me, *Cole*," he said cordially, struggling to keep his eyes off her teeth.

Dorothy Greene pressed her wrinkled fingers to her bulging lips and giggled like a schoolgirl. Then, with her quaking hand clattering the silver ladle in the crystal punch bowl, she sloshed a Styrofoam cup full of punch and jutted it at Cole.

Next in the trio was Lilly Hoffman, who was more reserved. She spoke quietly, drawing Cole closer. The more he leaned forward, the softer she spoke. Her breath smelled like hot August roadkill. Lilly Hoffman took Cole's hand and pulled him dreadfully close. She smiled and whispered vague things that Cole couldn't understand, the whole time patting his hand and blowing her rotten breath into his face. He held his breath, answering her with nods.

Cole was near fainting when Lilly Hoffman released his hand and gave him a fork and plate. He thanked her, then gasped for air over his shoulder and quickly stepped to the next lady, Helga Broadenhoff.

Helga Broadenhoff shook Cole's hand and nearly broke it. She was well over six feet tall and took up the entire side of the table. With her broad shoulders and hips, Helga Broadenhoff would have been Cole's first pick for a rugby team. She stood silent for a moment studying Cole's meek frame and said that he was too skinny to be a fireman. She heaped his plate with pastries, ignoring his pleas for moderation. She would not—could not—be ignored.

"Oh, Captain Walker, this is so exciting," she said, patting her chest like she was going to faint. Cole worried that if Helga Broadenhoff did faint, she might fall forward, toward him, on him. He sidestepped out of her impact zone. As she piled food on his plate with one hand, she ate a piece of something with the other. When she spoke, she sprayed projectiles of food over the array of refreshments.

"You must be starving. Try the strudel. I made it myself." Helga Broadenhoff slapped a humongous slab of pastry on the top of the heaped and sagging paper plate and thrust it at Cole. He took the plate and forced a smile, happy to be dismissed. He turned from the refreshment table and looked for a place to sit, preferably by a trashcan.

Cole found an empty folding chair in the back of the apparatus room where he sat, dissecting the oozing mass of pastries. He paused for a moment, cocking his head to the side to catch a curious sound similar to the rhythmic cadence of a steam locomotive. To his alarm, the sound was not from steam pistons but from the thighs of Helga Broadenhoff as she and her grating support hose marched toward him. The sound grew louder and louder until Helga Broadenhoff stopped directly in front of him.

"The strudel is a recipe from the old country, only I add extra mozzarella to spruce it up a little," chirped Helga Broadenhoff. She stood like a monolith squarely in front of Cole's chair. There was no escape. "What do you think of the strudel?" asked Helga Broadenhoff in a sweet voice heavy with German accent.

The question caught Cole off-guard; he looked again at the strudel. It glistened with specks of something unknown and smelled like apple pizza. He seized in panic like a deer staring into the headlights of an oncoming truck, a very large truck. Helga Broadenhoff took a bite of the chocolate cake in her fist and inquired again, hawking a wad of cake onto his knee.

"I just sat down, Mrs. Broadenhoff. I haven't yet found a moment to enjoy it," said Cole, though he made no attempt to.

"Call me, Helga," said Mrs. Broadenhoff, spraying Cole's food with a mist of chocolate crumbs as she spoke. She stood patiently in front of Cole, blocking his escape, waiting for his critique of the strudel while she gorged on chocolate cake. The goober of glistening cake burned into Cole's knee. Reluctantly, he cut off a piece of the strudel, nearly breaking his plastic fork. He raised the sample past his lips and over his head to break the streamer of cheese.

"It's a feisty bugga," she chuckled, launching another piece of chocolate cake from her mouth. From the cake projectile's trajectory, Cole estimated it landed somewhere in his hair. Helga Broadenhoff made a grab for the string of cheese spanning between Cole's fork and plate. She jerked on the string and pulled the entire cheese cap off the clod of strudel. She chuckled, playfully scolded the cheese, then wound it up like a yo-yo and placed the entire wad back on the strudel.

"Thank you, Helga," said Cole, moving the fork closer to his lips, frantically looking for an out. His eyes stopped on the mayor. Mayor Jenkins stood alone in front of the station and looked nervously over his shoulder as he fiddled with the decorative streamers. Mrs. Broadenhoff turned to look.

While her vigilance was broken, Cole stabbed the fork into his cup of punch and stirred off the strudel then slipped the empty fork into his mouth.

"Mmmmmmm! A delightful combination, Mrs. Broadenhoff. You have undiscovered talent."

"Call me, Helga," she chuckled, causing her melon-sized breasts to battle for position. She turned militarily and shuffled back from whence she came. Calling over her shoulder, she said, "You will have to try the borscht."

Cole held his smile and remained seated until the sound of chaffing nylons faded into the murmuring crowd. He stood and nonchalantly dropped his wad of strudel into the trashcan onto a larger mound of strudel.

"I saw that."

Cole's heart jumped. His mind searched for an explanation as he turned and looked into the sky-blue eyes of Evelynn Dewitt.

"But I won't tell," she said.

Evelynn Dewitt was the great niece of Dorothy Greene and the youngest member of the Auxiliary. Evelynn was a comely young woman, who spent most of her time staring at her feet. She wore an unattractive, ankle length dress and spoke in monotone. Her straight, lifeless blond hair hung in her face, partially eclipsing her eyes. When Cole noticed Helga Broadenhoff returning through the crowd, he suggested they leave the apparatus room. Without waiting for an answer, Cole grabbed Evelynn's arm and hustled her into the kitchen.

The rest of the firefighters were already in the kitchen. They were familiar with the Brookfield Ladies Auxiliary and had learned to hide early. Their laughter abruptly stopped when Cole and Evelynn entered the room. The only sound was Bleve's toenails as he clicked across the tile floor to investigate the newcomers.

Bleve stood beside Evelynn, then rolled onto his back trusting his exposed belly to his new friend. Evelynn recognized the honor and gently scratched his tummy, behind his ears and around his neck. She found the tag attached to his collar, *BLEVE*. Evelynn held the tag for a moment and looked up puzzled. "The dog's name is *Bleve?*"

"That's pronounced 'Blevy', like 'Chevy'," said Putnam, turning from the refrigerator.

"What does *Bleve* mean?"

"Bleve stands for Boiling Liquid Expanding Vapor Explosion." Putnam took a modest step closer to Evelynn. "BLEVE is what happens when a sealed container of flammable liquid is heated to the point that it explodes."

"What an unusual name for a dog," said Evelynn. Turning and speaking to Bleve, whose heavy head was now resting on her lap, "Why did you pick that name?"

Sounding almost annoyed, Valdez answered her question. "About seven years ago, Bleve was orphaned when his owner's mobile home caught fire. Captain Hastings couldn't save the man, but he was able to pluck a young pup from the trailer window just before the propane tanks BLEVE'd leveling the coach and the ones around it."

Uncomfortable with the inevitable comparison between himself and Captain Hastings, Cole slipped out of the kitchen. As he eased the kitchen door open, he saw Helga Broadenhoff, with a large plate of something, searching through the crowd. He stepped back into the kitchen.

"Captain Hastings must have been a brave man," said Evelynn.

"He was," said Valdez. The other firefighters nodded in agreement. The conversation lapsed into a moment of silence for Captain Hastings.

Ding…

The firemen stood in unison. Cole shoved open the kitchen door, knocking Helga Broadenhoff and the plate of pastries to the floor.

Ddddrrrriiiinnnngggg!

The firemen and their dog clotted in the kitchen doorway. The guests backed against the walls as the firemen burst from the kitchen, stepping over Helga Broadenhoff and running to their apparatus. The room fell silent at the dispatcher's announcement.

"Engine Sixty-one, Engine Thirty, Engine Forty-five, Truck Forty-five, Squad Sixty-one. Commercial structure fire in Sixty-one's district. 12074 Mariposa. Cross street Divine. Time out— Sixteen o'four."

Cole ran to the fire engine, passing through the gauntlet of blurred faces. He kicked off his dress shoes and stepped into his new rubber boots. He pulled up his brand new, bright yellow turnout pants and slid his suspenders over his shoulders; he was glad he had practiced. Just like the more experienced firemen, he did it without looking.

Unknowingly, Cole had also pulled up a pair of pink panties along with his turnout pants. Apparently, to everybody but Cole, the panties had been pinned and folded into his turnout pants earlier in the day. The frilly waistband puffed with rows of pink lace. Black stenciling across the butt read, *LA's FINEST*.

Captain Cole Walker's mind whirled and his hands trembled as he contemplated a plethora of fire scenarios. Poised in his pink panties, Captain Walker officiously ordered the crowd to stand back. He was pleased with the command tone of his voice but was surprised at the crowd's joviality. He climbed onto the fire engine, flashing everybody his *LA's FINEST*.

The crowd pressed back; some applauded, some laughed, others covered their ears anticipating the sirens. Two clouds of diesel exhaust lofted through the room as the fire engine and paramedic truck roared to life. In the distance, through the windows of the rising bay doors, they saw a plume of smoke.

"That's on the border of Brookfield and Lakewood. We better hurry or Engine Forty-five will beat us in!" snarled Barker as he slid the old fire engine into gear.

For Cole, everything was moving in slow motion, surreal, like an old black and white movie. Fixated on the distant smoke, Cole was unaware of the cheering crowd or the colorful banners that draped across the bay opening. In particular, he failed to notice the sagging welcome sign that drooped below the bay door header.

The lighter and faster paramedic unit was first to leave the station, blasting their siren, adding to the excitement and frustration of Engineer Barker. The paramedic truck easily slipped out to the street, under the drooping decorations—but not so for Engine 61.

The nylon cord drooped just enough to be out of sight of the engineer and the captain, but low enough to lasso the rear amber lights on the fire engine. For a sixteen-ton fire engine and an adrenaline infused engineer, the extra tension was scarcely noticeable. Captain Walker mistook the shouts and screams of the spectators for ovation and cheers. He gave the crowd a crisp salute, then turned his attention to the distant smoke.

As Engine 61 roared down the street, the nylon rope of the welcome banner tightened. First, it snapped free from the overhead tape, then cinched its noose around the card tables. The punch table was the first casualty, smashing the crystal bowl and large coffee urn to the floor. Glass platters, silverware, pastries and china dishes slid off the next cantered table and crashed to the floor. Helga Broadenhoff unsuccessfully lunged for the mound of strudel as her table collapsed and flew down the street. Chief Pierce was tripped by the whipping rope and splashed into a puddle of fruit punch. The guests stood in disbelief as the string of card tables whipped out the door behind the fire engine.

The distant column of smoke mesmerized Walker. He fumbled with the hand radios, fire hood, helmet and map. His mind spun wildly with plans for his first fire attack. The nylon welcome banner also spun wildly and coiled around the fire engine axel and brakes, then cinched tight. Engine 61 locked- up its rear axle and skid to a stop in the intersection of Main Street and Brookfield Avenue. Captain Cole Walker watched helplessly as the broken tables, the tattered welcome banner and the chance to make a good impression passed him by.

CHAPTER 4

SIDE TRACKED

Their Code-R response to Brookfield's commercial district was a long drive. The burning warehouse was on the far side of town, beyond the freeway and the railroad tracks. The firemen were further distanced from the warehouse because they had to stop and untangle the nest of nylon rope binding their axel, brakes and other parts of the fire engine's undercarriage.

Cole felt excited and nervous and embarrassed as he trotted around the crippled fire engine, with the red lights flashing, flipping the nylon cord off the light bar—all the while watching the column of smoke grow higher and wider.

"Need any help?" asked puzzled motorists as they slowly drove by the firemen, one in pink panties.

"Nope," Cole answered curtly over his shoulder, too busy pulling the cord wedged under the light bar. "The fire department is supposed to render aid, not receive it," he murmured.

"Do you guys know there's a fire over there?" asked another motorist, pointing to the smoke on the horizon. Engineer Barker did not even have the courtesy to return an indignant sneer. He slid underneath the rear axle and yanked at the frayed welcome banner stubbornly belayed around the brake hubs. Barker cussed and pulled, pulled and cussed, cussed and cussed. Bleve howled from the engine cowling. Cole's attention was divided between the hobbled fire engine and the ominous smoke plume. Sirens in the distance grew louder as they grew closer. Two fire engines sped through the intersection, the firemen waving as they passed.

Cole thought of his last sight of Chief Pierce. He was standing behind the engine, shouting, "Watch out!" Cole had thought the Chief was referring to the pending fire danger. Cole had put his index finger and thumb together and confidently signaled, *Okay, not to worry*. Now, a mile from the fire station and two miles from the fire, he was pulling welcome banners from the axle and card tables from the rear duals.

The radio blared. Engine 45 arrived first at the fire; Cole's fire. The captain reported a well-involved commercial warehouse with adjacent structures threatened on the south and east flanks. Engine 45 began the fire attack and passed command of the fire to Truck 45. The fire went to a third alarm.

Putnam joined Cole and together they jerked on the banner until it snapped free sending Cole to the ground and Putnam onto Cole. Finally, the last of the red, white, blue and blackened banner was unraveled from the fire engine. The three firemen and the howling dog climbed aboard the fire engine and once again, sped off to the fire.

"Looks like the recycling center," said Barker, studying the smoke. The billowing smoke had grown to a block wide. Large bursts of orange flame leapt into the air. They were only blocks away from the fire when the last of the third alarm units reported on scene.

The captain on Truck 45 was the Incident Commander, directing the fire attack and giving unit assignments. Cole waited impatiently for a clear moment on the radio to request an assignment. The Incident Commander did not acknowledge his request when he was one minute out. There was no response when he was 30 seconds out. Radio traffic was too heavy to talk, so Cole decided to get his assignment face-to-face from the I.C.

Boom! A fireball rocketed through the column of smoke, casting flaming debris in all directions.

"Probably one of their propane forklifts BLEVE'd," said Barker, not taking his eyes off the road.

When they turned onto Mariposa Street the blazing warehouse came into full view. *Brookfield Recycling Center* was painted in large, peeling letters across the warehouse's corrugated steel roof. The fire's intense heat obliterated the front half of the lettering, turning the metal roof purple and cherry red as if it lay over a welder's torch.

When Engine 61 finally arrived at the old, corrugated steel warehouse, Engine 45 and many other fire units had taken strategic positions around the structure. More than twelve fire engines encircled the building. Four ladder trucks were shooting water from their one-hundred-foot ladder towers.

When Cole hopped from the engine cab, he was struck by the noise and intensity of the fire. It roared and snapped and banged as items unseen fell to the inferno. From across the large open parking lot, he felt the heat bite against his bare cheeks. An overhead power line arced and exploded in an ardent flash, then fell to the ground sparking and jumping on the wet pavement. Cole ran to Truck 45, the Command Post, and pounded on the captain's door. The window rolled down, Captain Decker looked up from his fire attack plans and down on Cole.

"Welcome to the party, Walker," said Captain Decker, with dour expression. His eyes burned into Cole with more intensity than the fire behind him. "Isn't this *your* district?" sneered Decker, with his damning, rhetorical question.

Cole felt his blood pump to his face. Explaining his delay to Captain Decker would sound too much like groveling, and he did not want to grovel—not to Decker.

Cole avoided the question. "What's our assignment?" he shouted, trying to yell over the roar of the fire.

A thunderous explosion rumbled from the fire, distracting both their attention. A churning ball of fire shot skyward as another propane tank BLEVE'd. The force of the blast toppled several stacks of flaming cotton bales, sending firemen retreating to safer positions.

An eight-foot section of metal roofing launched from the explosion, spiraled through the air and slapped the ground on the other side of the command post. Captain Decker watched the red hot metal until it stopped sliding before returning his attention to his worksheet. He studied his assignment roster, tapping the clipboard with his pen.

"Looks like the most critical assignments have been filled by engine companies that arrived before you." Annoyingly, Decker resumed tapping his pen against the clipboard, studying the hastily drawn schematic of the warehouse and the surrounding buildings. Then, thoughtfully changing positions, he pensively tapped his pen against his pursed lips. "Here!" said Decker, stabbing his pen onto the paper.

"Here's a good spot for you," said Decker, delighted. "Exposure protection," he said. "In the rear. Try to get there before the fire goes out."

Turning his back to Cole, Captain Decker rolled up the window as he picked up the radio microphone and demanded status reports from the fire ground units. Cole backed away from the ladder truck and stood defiantly in his pink panties. Captain Decker was too busy to notice Cole's obstinacy, so Cole held his position and watched the fire in protest.

The large steel warehouse was stacked from floor to roof with tightly bound bales of rags, half of which were on fire. On the west corner, a crew of firefighters played their hose lines into the angry flames. Like a crazed, caged animal, the fire lashed at the firemen with claws of flame. Clouds of steam exploded, dwarfing the firemen as their water streams bored into the heart of the flames.

Cole figured enough time had lapsed for Captain Decker to have finished collecting his status reports. He rehearsed what he was going to say to Decker one more time before turning around to find Captain Decker standing directly behind him. Cole gasped aloud.

Captain Decker stood rigid, sporting a crooked grin. He studied Cole, enjoying his moment of surprise. He was not sure which he enjoyed more: Cole's startled reaction, or his pink panties. Captain Decker's smirk widened to a devilish smile. The fire illuminated his face, causing the shadows of his sharp nose and sunken cheeks to move and flicker. The fire danced in his loathsome eyes. "Walker, your assignment is exposure protection on the south side. The *back* of the warehouse. Get to it."

There are plenty of areas that need more immediate attention, thought Cole. He tried to explain his strategy to Captain Decker. "We could help Engine Forty-five advance their line…"

"Contain…" interrupted Captain Decker, "then extinguish." Decker's overbearing demeanor was reminiscent of the fire academy. Cole stood defiantly, but silent, just like in the academy. He tried again to reason with Decker and get an active firefighting assignment.

"If we position ourselves between the northern units, we could hold the fire…"

Captain Decker held up his hand and waved it side-to-side, stopping Cole mid-sentence. Decker stepped forward, closing the gap between the two men. "If you want to command the fire, be first in. You have your assignment," snapped Decker and returned to the truck.

"Do you have an assignment for Squad Sixty-one?" Cole called to Decker's back.

"You keep them," Decker yelled over his shoulder.

Cole squeezed his clenched fists until they hurt as he marched to the engine and relayed Decker's orders to the crew.

"Exposure protection!" shouted Barker and Putnam in unison. Cole did not respond. Barker huffed in disgust, then pushed the shift lever forward and headed to the rear of the warehouse.

Though important, exposure protection was a tedious job on the outer fringe of the more exciting firefighting operations. To the rear of the Brookfield Recycling Center were exposures of only modest value: a couple of sun-bleached rag bales, a cannibalized forklift and two small outbuildings housing a small clutch of empty propane tanks.

The objective of their assignment was to shoot a heavy water stream between the fire and the threatened structures forming a cool, protective water curtain. Unfortunately, immediately behind the building were the Metrolink tracks that did not allow adequate room for their operations. Cole decided it was safest to set up their defensive position south of the railroad tracks.

In only three minutes, the crew had the four-inch supply line laid and connected to Engine 61's pump intake. Barker throttled up the diesel motor. The deck gun mounted atop the fire engine sputtered and spit before shooting a solid two-inch stream over the tracks, between the burning warehouse and the uninvolved structures.

Captain Decker made a radio report declaring the building unsafe and its contents unsalvageable. All fire ground units were ordered out of the building and into defensive positions. "Surround and drown" meant a protracted ordeal. Cole and his crew settled in for a long siege. They watched the dark figures of firemen, silhouetted against the goliath flames, struggle with their uncooperative hose lines. The firemen dragged their heavy lines into position and braced against the heat until the water spray darkened the fire, forcing the flames to reappear around another cotton bale. Then, the firemen would readjust their position and repeat the scenario.

The news reporters were just as relentless as the fire. Shiny vans with telescoping microwave antennas descended on the warehouse. Halogen camera lights illuminated the slew of reporters as they primped and posed in front of the inferno. Also, Public Information Officer Red Hall spun into the parking lot in his shiny PIO sedan, bumping the trash dumpster and sending it whirling into the herd of media. Red undraped his pressed, bright yellow turnout coat from under its dry cleaner plastic and slipped the coat over his white PIO shirt. The reporters and the camera crews swarmed to him.

Cole caught a narrow glimpse of Red Hall between the buildings and wondered if Red had gathered enough information to satisfy the throngs of reporters. Suddenly illuminated by the burning lights of the news cameras, Red Hall seemed to glow. He fielded their questions until upstaged by an exploding propane tank. Just as quickly as they flooded to him, the reporters receded, opting for more dramatic footage.

As the fire continued to hide from the boring hose streams, the battle looked like a stalemate. The heavy, frigid night air descended thick over the fire ground. The misting over-spray from the elevated water cannons was biting cold. The bitter night proved too much for the news crews. The camera lights eventually shut off, the telescopic antennas retracted, and the news vans drove away. Cole buttoned the collar to his turnout coat and tried to get comfortable in the upright, rigid, uncomfortable captain's seat.

With the reporters gone and the fire attack in defensive mode, the thrill and novelty of the fire faded to routine, then boredom. Midnight dragged into two o'clock. Two o'clock slipped into four in the morning. For Cole, the exhilaration of his first fire had long since worn off. Earlier, during the excitement, he had released the paramedics to return to quarters and volunteered to keep watch while Barker and Putnam slept. Now, he had a hard time staying awake.

A little before 5:15 a.m., a beige sedan honked. About a halfblock away, Helga Broadenhoff and the other ladies of the Ladies Fire Auxiliary squeezed out the car along with several cups of hot coffee. They set the Styrofoam cups and a box of donuts on the milk crates they used as make-shift tables, then waved the firemen over.

Cole woke Barker and Putnam. Since the water cannon was in the same position it had been in all night, Cole figured it didn't need monitoring, so all three firemen headed for the coffee. They walked clumsily along the uneven gravel of the train tracks. Their joints and backs were achy and stiff from sleeping in sitting positions.

Helga Broadenhoff held out three large coffee cups like beer steins. The coffee tasted good and warm. Lilly Hoffman, preceded by her breath, catered to the firemen offering them selections from the donut box. The frozen firemen ate donuts and stomped their feet, trying to shake off the morning chill. From a distance, they all huddled and watched as Engine 61 droned relentlessly, hurling tons of water across the train tracks to a fire that stubbornly held its grip of the warehouse.

In the early morning silence, Evelynn Dewitt studied Cole and asked if the pink panties were a tradition. Cole lifted his turnout coat and stared in shock at the panties clinging to his waist. He dropped his coffee cup and jumped about like his pants were on fire, wiggling and wriggling out of the grip of the support panties pinned to his waistband. His crazed antics were disrupted, shattered by the blast of an air horn as a train came into view. It was the Tuesday morning, high speed Metrolink loaded with commuters.

The seven stood numb and motionless. The firemen and the ladies of the fire auxiliary stared at the approaching train, then back at the two-inch water stream shooting directly across the train tracks. Instantaneously, as if starting a race, the firemen dropped their donuts and ran toward the fire engine. The train had a head start. Before they could cover half the distance, the train broke through the water stream.

From where Cole stood, panties bunched around his ankles, he could see the train passengers in striking detail. They were all pressed against the opposite windows of the passenger car, straining to glimpse the burning warehouse. Shackled by panties, Cole watched helplessly as the locomotive cut through the water stream with a slap. The hammering water thundered against the solid metal locomotive housing, then sizzled through the generator grates, pinged against the handrail, and crashed through the engineer's window. There was a brief silence as the stream shot unobstructed between the locomotive and passenger cars. Then, just a quickly, the powerful water stream resumed its rhythmic slapping, like the sound of a child dragging a stick along a picket fence, but much, much louder.

The two-inch stream blasted into the passenger cars hitting window, stile, window, stile and so on. The water uniformly slapped down each pane of Plexiglass, so the 2,000-gpm water stream was unobstructed to deluge the unsuspecting passengers as they clustered against the far side of the railcar.

Reaching the fire engine first, Putnam climbed on top and struggled to turn off the water valve against the pressure of the pump. Reaching the pump control panel moments behind Putnam,

Barker punched the throttle panic button dropping the water stream to a dribble just as the last railcar passed. Captain Cole Walker, realizing the futility of his run, slowed to a hobble. With panties bunched around his ankles, he watched in disbelief as the train full of passengers and water disappeared down the track.

CHAPTER 5

BABY SHOWER

Captain Cole Walker's station assignment was his first experience with the firefighter's work schedule. The 48-hour break that followed his first grueling 24-hour shift was a welcome relief. The weight and flashbacks of his first shift's fiascos dominated his thoughts, preempting his motivation and desire to do anything constructive. Cole used the 48 hours to recuperate and regroup.

For two days, Cole poured over his training manuals seeking guidance, but found nothing related to water monitors vs. trains, or how to recover from destroying an open house celebration. He eventually rationalized that the destruction of the Brookfield Ladies Fire Auxiliary's equipment, wearing women's underwear to a third alarm fire, and deluging a passenger train were all innocent mishaps to which any fire captain would have fallen prey. The calamity of his first shift was nearly forgotten over his two days off and Cole was hopeful everybody else had forgotten it, too—until he saw the fire station.

Arriving several hours before shift change, both his jaw and shoulders dropped as the firehouse came into view. Slamming the transmission into Park and jerking his car to a stop, he slid to the passenger window and stared at the banner.

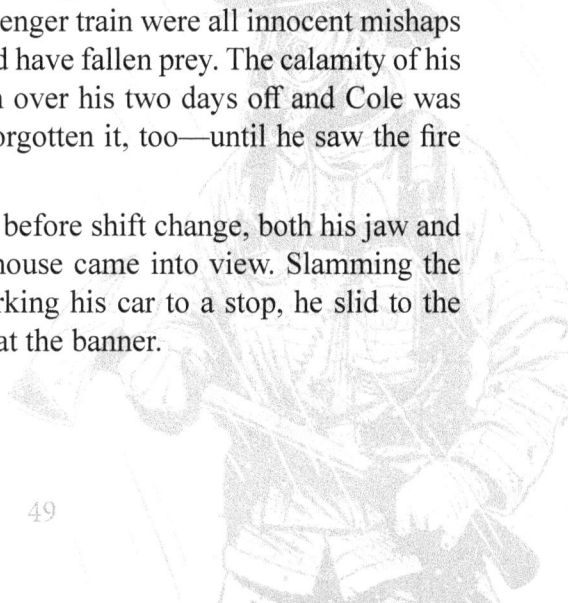

Draped across the front of the fire station as an artist's canvas was the 16- foot salvage cover, depicting a train half-submerged in water. *Fire Station and Train Wash* was scrolled across the top in bold black letters. The sign was too high for Engine 61's ground ladders. Only a ladder truck like Truck 45—Decker's truck company—could have raised the banner. Cole's car tires squealed and smoked as he spun into the parking lot.

The off-going firefighters were also surprised to see the train sign. One-by-one, they collected in front of the station to gawk at the anomaly. A half- dozen firefighters in various stages of dress fanned across the front apron reading and re-reading the sign. Some whispered together while others boldly laughed aloud. Cole felt the weight of their stares. Already, one hour before his second shift began, the pressure, humiliation and insecurity returned.

Cole tried to occupy his tailspin thoughts by wiling the early morning planning the day's schedule. He read and re-read the department's protocol on morning line-ups. The procedure was simple. At precisely 0800 hours, he would announce the line-up on the station's intercom. The crew would then line up shoulder to shoulder on the apparatus floor between the fire engine and the paramedic truck. He would then read notes from his daily planner and outline the day's activities. *Simple.* Cole reminded himself that there was an intricate balance in converting these provincial, small-town firefighters to unfamiliar Los Angeles Fire Department procedures. He realized his crew had suffered the tragic loss of Captain Hastings and were sensitive to new leadership and change. He would allow the firemen time to adjust until they were comfortable with him and the new ways.

At nearly eight o'clock, Cole looked around the office for a public- address microphone. Searching futilely, he discovered that the antiquated fire station didn't have a PA system. Instantly, he created Plan B.

All the firemen were in the kitchen. Cole stood in the apparatus room just outside the kitchen door and checked his watch. From the kitchen, he heard bursts of laughter separated by periods of hushed conversation, then more laughter. He pressed his ear against the crack of the door but couldn't decipher any of the conversation. He checked his watch again. The time was eight o'clock straight up.

"Attention C-shift. Fall in for morning line-up on the apparatus floor," said Cole from the crack of the kitchen door. The firemen stopped their conversations, stared curiously at the sliver of their captain's face in the door crack, then back at each other. In silence, they slid their chairs back and shuffled into the apparatus room.

"Come on, Putz. That means you, too," said Valdez, as he knocked on the open refrigerator door. Putnam returned the plates of leftovers but kept the chicken thigh clenched in his teeth.

Once the crew was in line, Captain Walker faced the men and inspected their formation. Barker and Valdez stood with slumped posture, while Putnam and Razmunson stood with over-exaggerated stiffness. Putnam was in uniform with chicken flesh protruding from his lips. Raz and Val were in exercise shorts and tee-shirts. Barker was unshaven and still wearing civilian clothes.

Cole raised his clipboard to commence his first official line-up. Engineer Barker cleared his throat. Cole looked up from his notes and noticed Barker was board-stiff and raising his hand.

"Yes?" asked Cole.

"What is the purpose of a line-up, sir?" asked Barker, slowly lowering his hand in a saucy gesture of obedience.

Captain Cole Walker spewed the answer verbatim. "A line-up is a time to disseminate information and outline the day's activities," he quoted proudly.

Again, Barker raised his hand only shoulder high, fanning his fingertips in a sarcastic gesture, "Why can't we disseminate information while sitting comfortably around the kitchen table?"

"Well…" Cole said, seized in mid-thought. "I…I guess there wouldn't be any problem with that…" His voice trailed off as he watched his crew break formation and parade back into the kitchen.

There had never been a question posed like that as far as he knew. In fact, he wasn't even sure if this was the way other stations conducted their morning line-ups because, up until today, he had never been involved in a station line-up. He clutched his clipboard against his chest and draggled behind the crew back to the kitchen.

The four firemen were quick to reenter the kitchen. The swinging kitchen door flapped back and forth in front of Cole, making it difficult to gracefully time his entrance. The door slapped hard against his hand, a result of his bad timing. When he pushed the door open, Barker, Val and Raz were sitting at the dining table, cradling their mugs of coffee. Putnam had resumed his position behind the refrigerator door.

Cole walked in last and upon his entrance, the room fell silent as the crew picked up their coffee mugs and sipped in unison. Cole felt naked and conspicuous. He thought a cup of coffee would help him assimilate with the crew. His clipboard made an obtrusive clatter when he set it on the table, then he pulled open two or three cabinet doors before finding the cupboard with the coffee mugs inside. He grabbed the black mug with *Captain* printed on it and filled it with coffee. He looked curiously at the can of black boot polish on the counter but dismissed the thought as he took his seat at the head of the table.

As Cole sipped from his mug, he tried to break the uncomfortable silence. "Mmmmmm. Good coffee. Gourmet blend?" he asked, his lower lip eclipsed with a black smudge.

Cole noticed that none of the firemen could hold eye contact with him for more than two seconds—an obvious sign of respect for his rank. His confidence grew as fast as the smudge on his mouth widened. Each sip of coffee spread boot polish toward the corners of his mouth, eventually developing into a large, hideous, ebon grin.

"Did you see the news of the fire?" asked Putnam, peering over the refrigerator door. The condiment jars softly tinkled from his quaking as he struggled to subdue his laughter.

"No. I missed the news," smiled Cole, spreading a friendly, black grin.

"I have it on disc," said Putnam, "Wanna see it?"

"Absolutely," said Cole.

Putnam sprang from behind the refrigerator door, wedged his way between the semicircle of recliners at the far end of the room and slid a DVD into the player. He fumbled with his thick fingers on the small buttons of the remote control, fast-forwarding through the commercials then stopping on TV-3's Jennifer Preene posed in front of the Brookfield Recycling Center.

The news camera showed flames coming out of every opening and firemen battling the fire. The scene was very dramatic. The beautiful reporter speculated on the cause of the fire and extinguishment efforts. Loaded with questions, she turned to the Los Angeles Fire Department's Public Information Officer.

"Shhhhh!" said Putnam, pointing to the television, "This is the good part."

Coming into view with the panning camera angle was Red Hall. Apparently, he did not know he was in the camera's field of view as he stroked his tight wavy, red hair. He picked at something in his ear and then studied his finger.

"Uuuooooo!" shouted the firemen, crescendoing into laughter.

"Shhhhhhhh!" said Putnam, pointing to the television.

PIO Officer Red Hall stepped next to Jennifer Preene and gave a nervous, waist-high wave. The right collar of his bright yellow turnout coat was folded under, causing the left collar to stick straight up. Jennifer Preene held her microphone to the fire department's representative. Before speaking, Red pulled on the microphone, bumping it against his mouth. He tried again to commandeer the microphone, but the reporter would not relinquish it. When the reporter unexpectedly released the mic, Red hit his helmet, knocking it askew.

Jennifer Preene snatched her microphone from Red Hall and began her interview by asking how many fire engines were fighting the fire. Red Hall turned his back to the camera and counted the units he could see. "At least six," he answered.

"How long until the fire is contained?"

Red didn't know.

"How did the fire start?"

Red shrugged. Then he waved *Hi* to his mom. The camera whipped around to catch an exploding propane tank abruptly ending the interview. In concert, the entire crew exploded in laughter—except Cole, whose cheeks burned hot with flush. It was another example of L.A.'s finest.

Putnam rewound the disc, stopping short of the fire. He replayed the interview three or four-zillion times before returning to the refrigerator. Inspired by the news footage of the fire, the line-up conversation deteriorated into the swapping of war stories. In a game of one-ups-manship, the firemen spun their yarns of big fires and narrow escapes.

Valdez started first, telling about the time when he was almost crushed when a two-thousand-pound industrial air conditioner fell through a burning warehouse roof, narrowly missing him. Razmunson jumped in next, saying he was almost electrocuted when his air bottle arced against an electrical panel. Barker wove his story of the time when the wall of a burning industrial tilt-up fell over him. Instinctively, he dove through the loading door as the wall collapsed around him.

"The first story never stands a chance," conceded Valdez, which was true. Each tale sparked relics of more astonishing stories, each more elaborate than the one before. The stories, compounding one upon another, were raising the volume until the kitchen was awash in indecipherable racket. Engineer Barker silenced the crew with a wave of his hand and said, "How about you, Cap. Let's hear one of your stories. Any close calls?"

The room fell silent as each fireman leaned forward to hear the fantastic chronicles of their new fire captain.

"Oh, that's alright," said Cole. He waved his hands like a celebrity trying to quell applause. "I don't need to interrupt you guys. Continue. Please. I was enjoying your stories." Cole took a long sip from his coffee mug, while he surveyed the other firemen over the rim.

"Were you on the Malibu brush fire of '93?" asked Putnam, with sincere curiosity.

Cole innocently grinned with his blackened lips and thoughtfully set his black mug on the table."Nope."

"Were you on the Lakewood Mall tunnel fire last year?"

"Nope."

"Have you ever worked on the fire boat?"

"Nope."

"What was your assignment before you came here to Brookfield?" probed Valdez, suspiciously.

"I was the Public Information Officer," Cole said, happy to finally have an answer, even though it sounded vague.

"Sounds like you had to do time in headquarters before you got promoted," said Barker. "What did you do before that?"

"I was assigned to the Petroleum Storage Systems Bureau in headquarters," said Cole. He didn't like where the interrogation was going but couldn't change its direction without raising suspicions.

"How about before that?" asked Razmunson.

"Training Manual Revision."

"Before that?"

"Procurement." Cole felt his cheeks warming from blush and tried to hide his reddened face by sipping from his coffee mug. He felt his heart pounding in his ears.

The room was silent for a brief moment, then came the most damning question of all, courtesy of Richard Valdez.

"Have you ever worked in the field?"

"Besides this station?" asked Cole, groping for humor. He pensively raised his coffee to his blackened lips. "Nope."

"This is your first assignment in a station?" asked Putnam, unable to hide the alarm in his voice. Valdez let out an unintentional huff and scanned the faces of the other firemen. Synchronized, the entire crew slowly sipped their coffee.

Cole tried to regain his composure by offering a feeble explanation about the value of his experience in PIO and the Petroleum Storage Bureau. "Don't underestimate the value of books and watching upper management in motion," Cole said, not believing it himself. The firemen sat quietly, deliberately sipping from their mugs and exchanging glances with one another.

Following the line-up and his humiliating revelation on job experience, Cole withdrew to the security of the captain's office. He flopped lifelessly at his desk. As he slowly revolved in the swivel chair, he scanned the office walls and studied the pictures of large fires, horrendous traffic accidents, firemen shaking hands with Mayor Jenkins, and other disasters.

As he turned, the room spun dizzily. Pictures of tremendous car wrecks and burned buildings blended one into another. Cole stopped his spin in front of the larger-than-life portrait of Captain Keith Hastings. Captain Hastings was staring down at him. He always stared. No matter where Cole was in the office, Captain Hastings stared at him. Cole moved his head to the right, then to the left. Captain Hastings' penetrating stare never left him. Cole raised his head higher, then lower, then he crouched with his eyes just above the desktop, but Captain Hastings' stare still bored into him. Cole leaned forward, then back. Nothing could break Hastings' stare. Cole turned his back, then suddenly looked over his shoulder - or at least that's what he was in the process of doing when he noticed Barker standing in the doorway.

"What can I do for you, Rus?" asked Cole, nonchalantly.

"Well, Cap," Barker started, stepping forward into the room and pulling the door closed behind him. "I was wondering if I could share some thoughts with you."

Cole nodded and took another sip of coffee. The smudge of boot polish covered his lips, both cheeks and most of his chin.

Barker uncomfortably patted his hand over the few strands of "hair glued to the top of his obviously bald head. "Cap, I only have 18 months until I retire. Heaven knows, I've seen more tragedy than I signed on for. No offense, sir, but in your situation, you're in the position to put the crew in harm's way. And… I… well, I don't know the politics involved with this annexation, but I hope they don't take priority over the safety of the men." Barker gave a quick nod, then as quietly as he came in, he left.

Cole sank heavy into his chair. *Barker is right*. He was in the position to get his men hurt. This was not a game. *Maybe I should talk to Chief Pierce about transferring back to headquarters?* Cole's thoughts were interrupted when Barker softly knocked and stuck his head through the partially opened door. "You have something on your lip," he said, uncomfortably, and then made rubbing motions to his own lips before pointing to Cole's face. Obviously uncomfortable with his gesturing, Barker exited.

Turning to Captain Hastings, Cole asked, "Do you have comments, too?" Reflected in the glass of Hastings' portrait Cole saw his own blackened face. He touched his chin and examined his smudged fingers for a considerable time before realizing they were covered with boot polish. He fell into his chair defeated and lowered his head to the cool glass desktop. The open house fiasco and the train incident came crashing down upon him. He knew textbook tactics and strategy, but the onslaught of recent events made it painfully clear that this assignment was too much for him to handle. He decided to call Chief Pierce and request a transfer back to headquarters as soon as he could clear the lump in his throat.

Rrrrriiiinnnngggg!

Startled by the telephone, Cole snapped upright. He cleared his throat and spoke aloud to hear if his voice betrayed any sign of distress. The phone rang again. Cole cleared his throat again and picked up the receiver. "Fire Station Sixty-one, Captain Walker."

Sandra Goodale was on the line. Her voice soothed him. Disregarding the shoe polish on his face, Cole used both hands to cradle the phone to his ear.

"Sandra, it's good to hear from you. How are you?"

"Fine, Cole. I hear it's pretty active down there."

Cole knew Sandra was referring to the disastrous open house and the train incident. "Oh, Sandra," he said, as he swiped his chin with a piece of typing paper. "You don't know the half of it."

"Don't let them intimidate you," she consoled. "You just be yourself, Cole Walker, because that's what you do best." Her advice carried the maternal timbre that he had always liked about her. She continued, "Hold on. The Chief wants to talk to you."

Cole held the phone to his ear and leaned to the picture of Hastings to check his chin. The smudge had lightened but was now smeared wide across his face. He threw the blacked paper into the trash and grabbed another sheet.

"Captain Walker," the Chief's voice boomed through the earpiece. Chief Pierce continued without allowing Cole a chance to reply. The Chief's voice rose steadily in volume. "I was hoping you could explain why Metrolink sent me one bill for train repair and another bill for dry cleaning 28 business suits, three dresses, two habits and a sari!" His sentence crescendoed into a yell.

Cole jerked the receiver from his ear as Chief Pierce continued."Perhaps they didn't teach you in the fire academy, Captain Walker, that WE DO NOT SHOOT TRAINS!"

Cole held the phone at arm's length, waiting for a moment of silence before attempting to speak. He cautiously pulled the receiver to his mouth. "Well, Chief, I was going to call you regarding that..."

The Chief interrupted, which was okay with Cole because the lump in his throat had returned. "Mayor Jerk-ins is coming to your station today. He wants to see your operation firsthand."

"Mayor Jenkins... here? Today?"

"Right," said the Chief. "So let him ride along on the engine. Show him the nuts and bolts of our operation."

Cole sedately acknowledged the Chief. If today was going to be his last day, he certainly did not want to spend it kowtowing to Mayor Jenkins.

Again, Cole started to ask for a transfer, but was again interrupted by Chief Pierce.

"I also got a call from Captain Decker, Truck Forty-five. You know him, right?"

"Yes," sighed Cole. "I know Captain Decker." Cole thought of the train banner hung on the firehouse. "What did *he* want?" asked Cole, allowing his agitation to pierce his voice.

"Decker thinks the Brookfield annexation might be too much for you. He suggested that he should take your place because you can't handle it." The Chief gave a dramatic pause, then asked, "Do you want out?"

The phone went silent as Chief Pierce awaited an answer. Cole imagined Captain Decker taking command of *his* firehouse. The thought made him bristle.

"It's no problem, Chief. I can handle it," Cole blurted, hoping to muster enough confidence in his voice to believe it himself. There was no lump in his throat.

"Okay, Captain Walker, you can stay, but let me give you some advice."

Cole detected a hint of compassion in the Chief's voice. He pressed the phone tight to his ear.

"DON'T SHOOT ANY MORE TRAINS!"

The phone line went dead.

"Thank you, sir. I will use that advice right away, sir," said Cole to the dial tone. He laid the phone back on its cradle and his head back on the desk.

Cole finished cleaning his face, using a half ream of typing paper, then he went outside to tell the crew about the mayor's visit. The firemen were in the rear parking lot, rolling fire hose.

Cole made his announcement, which caused the crew to moan and grumble. Apparently, they did not like the mayor any more than Cole. Cole asked Val to plan on the mayor for lunch.

"He better coughs up four bucks for chow," snapped Valdez. "He stiffed me last time."

Cole remained in the back lot to watch his crew roll 50-foot lengths of fire hose into storage rolls. Valdez dropped the dry cotton fire hose from the 30-foot hose tower. Razmunson and Putnam rolled the hose into tight rolls, while Barker lubricated the swivel couplings with graphite. Barker's latex gloves were blackened from the graphite powder. Cole picked up the quart can of the dry lubricant and examined the graphite shimmering like ground pencil lead. Barker cautioned him that graphite was easily airborne and adheres to anything it touches. Without looking up, he flashed his gloves as evidence. Cole blew across the open can, causing a gray cloud of graphite to loft into the air and settle on the top of Barker's bald head. With each tilt of Barker's head, the graphite flakes shimmered and sparkled in the sunlight. Cole returned the graphite can to Barker and returned to his office to evolve a plan.

"Oooo's" and "Ahhh's" filled the room when she held up the large stuffed teddy bear, the last gift of the baby shower. The new mother-to-be scooted to the edge of the couch and awkwardly stood up. She stepped into bunny slippers, her feet too swollen for shoes. No one would have thought such a spirited girl would domesticate so quickly; first the pregnancy, then the marriage, then the bunny slippers.

"Time for cake," announced the hostess, directing the herd of women to the dining room for dessert and palaver.

"I've still got the moves," said the pregnant woman as she waddled across the living room. Her actions did not betray her feelings. She hated what this pregnancy was doing to her shape, as well as her lifestyle.

Too pompous to knock at the front door, Mayor Jenkins entered the fire station unannounced through the rear door and strutted directly to the captain's office. He opened the door without knocking, catching Cole doodling on a scratch pad.

"I hope I'm not interrupting anything," said Mayor Jenkins, knowing that he wasn't.

Startled, Cole slid his doodle pad under a pile of forms and welcomed the mayor, ignoring his arrogance. Cole was first to force a smile and extend his hand. The two men shook hands with no warmth or vigor—once up and once down, with a clean break, the kind of handshake wrestlers exchange before a match. Mayor Jenkins strolled around the office like he owned it.

"Perhaps you'd be more comfortable in the dayroom, Mayor," said Cole, diplomatically. "We never know how long it will be before our next bell."

The mayor's feigned smile quickly changed to a somber countenance, thinly veiling his contempt. "Captain Walker, *I am* familiar with fire department routine. Quite frankly, I am here to investigate *your* routine. I'm sure you are aware that I opposed LA's annexation of my fire Department. I think it's a grave mistake."

"Well, Mayor," said Cole, curt but civil, "I hope to prove you wrong." Cole held the office door open for the mayor and escorted him to the kitchen. "There's fresh coffee in the pot and newspapers on the table," said Cole, trying hard to appease the mayor.

Mayor Jenkins snatched the stack of newspapers and sauntered to the recliner chairs at the far end of the dayroom. He pulled a pack of cigarettes and a lighter from his shirt pocket. He pulled a cigarette from the pack with his lips.

"Sorry, Mayor, this is a non-smoking building," said Cole.

"Is that so?" asked the mayor, flicking the lighter as he raised it to his cigarette. He casually drew from the cigarette, bending the flame to the cigarette until its tip glowed red. Mayor Jenkins blew a rack of smoke at Cole and apathetically sat in a recliner. In one fluid motion he jerked his hands apart, opening the newspaper with a snap.

Ding…

The station sprang to life. The firemen ran into the apparatus room from every direction and Bleve assumed his firedog position on the motor cowling between the rear seats. The mayor begrudgingly allowed Putnam to assist him to the jump seat. He sat nose to muzzle with Bleve who dropped his ears and growled.

As the alarm bell reverberated into silence, the dispatcher announced a "medical aid" at the legal offices just down the street from city hall.

Although the response time to the legal offices was only two minutes, in that time they were tailgated down Main Street, gutter-sniped at Elm and nearly T-boned by a woman painting her fingernails and steering with her elbows. Everyone was tense when they finally stopped in front of the six- story office building.

The two paramedics carried the IV box, trauma box, drug box and telemetry unit. Putnam carried the oxygen tank and an armload of cardboard splints. The six men and their equipment left scarcely any room in the cramped elevator. The stench of stale cigarette smoke and halitosis emitting from the mayor made the elevator's tight confines seem even smaller.

The elevator opened to the sixth floor and the firemen spilled out, grateful for the fresh air. Loaded with equipment, the firemen lumbered their way to the legal office and inquired about a victim.

"Here I am," waved a pretty, young secretary before returning her finger to her mouth. She pressed the telephone against her salacious breasts, muting it. "I cut my finger and it hurts like heck."

Valdez knelt at her side and examined the finger she held out at breast level. "You mean this scratch here?" he asked. "How did you cut it?"

"On that nasty thing," she said, pointing at the sheet of paper sticking out of the typewriter.

Valdez stood up and stepped back. "You called 9-1-1 for a paper cut?"

"I didn't have a band-aid," defended the woman.

"We almost got killed driving down here," added Razmunson.

"At what point in time did you feel that your life was in jeopardy?" interjected Captain Walker, not allowing her to answer. "Val and Raz, go available and return to quarters." The two paramedics gathered their equipment and left.

While Putnam gingerly put a bandage on her finger, Cole admonished the woman.

"9-1-1 is for real emergencies, ma'am. Please consider that before making an emergency call." He made no attempt to hide his irritation. He made an abrupt about-face, nearly colliding with the mayor, and marched from the office. Mayor Jenkins remained, consoling the woman. He gently examined her finger, then gave her a sympathetic hug. The woman pushed away, repulsed by the mayor's dandruff and nicotine breath.

Mayor Jenkins called to Captain Walker as he stomped down the hall. "You could be a little more considerate, Captain Walker. She is a taxpayer."

"I was considerate," said Cole, stopping and allowing the mayor to come face to face, "I didn't have her arrested." Cole turned and walked to the elevator. "You better pick up the pace, your lordship, unless you want to walk back to the station."

Inside the elevator, the air was drenched with anger, nicotine and the mayor's rancid breath. The ride back to the station was no better. After they backed into the apparatus bay, the mayor was waiting at Cole's door when he stepped off the fire engine. "I'd like to have a word with you, Captain Walker."

Hopping down from the engine cab, Cole leaned into the mayor's face, causing him to arch backwards, "Let's go to *my* office."

Mayor Jenkins paced the floor inside the small office, while he lectured Cole on professional courtesy and how to treat taxpayers who pay his salary. Cole retorted with his interpretation of arrogance and stupidity, and the wasting of government services. The mayor went for the throat. "Are you mad because she was cutting into your doodle time,Captain Walker?"

Cole stepped forward, reducing the gap between himself and the mayor. "We have an entire crew of men ready and willing to risk their lives to save somebody else's. Every time they're called away to put a bandage on a hangnail or check somebody's hemorrhoids, they can't protect those who need it most. Now excuse me. I have work to do," said Cole, as he stepped aside and pointed to the door.

The crew, pressed tight against the other side of the office door, scrambled back to the fire engine and began wiping it with their chamois. Mayor Jenkins held the doorknob and turned back to Cole, "Your time here is limited, Captain Walker." Jenkins shoved his way out of the door, passed the firemen as they merrily whistled and cleaned the spotless fire engine.

The house was filled with laughter, cackling women and forks clanking on empty cake plates. The young mother-to-be looked at her watch. "Oh, shoot! I'm late! I'm supposed to pick up Gary from the shipyard at 1:30." She gulped the last of her orange juice and, with considerable effort, pushed herself up from the table.

"Do you want me to call Gary and tell him you're running late?" asked the hostess.

"No, that's okay. I know a short cut," she said as she waddled to her mini-truck. In her sport truck, she didn't feel as pregnant because she had agility and speed. In her little truck, she wasn't hindered by her awkward size, although the seat belt hurt her stomach, so she didn't wear it.

The other women loaded the cab with the baby clothes, stuffed animals and leftover cake. Then they clustered around the driver's door, taking turns hugging and kissing the young mother-to-be. The woman waved good-bye as the rear tires chirped and she sped away to her next appointment.

There was plenty of food left over after lunch. Mayor Jenkins was a light eater, and so were the firemen today. The open doors and windows did little to alleviate the noxious odor that the mayor had brought to the kitchen. After eating, Valdez approached Mayor Jenkins and requested four dollars for the meal.

Ding...

Mayor Jenkins, conveniently distracted by the bell, returned his wallet to his pocket before paying for lunch. He energetically strode to the fire engine.

Dddddrrriiinnnggg!

"Engine Sixty-one. Squad Sixty-one. Traffic Collision. West bound Ninety-one Freeway, west of Brookfield. Reported persons trapped. Thirteen thirty-four."

Cole quickly hopped on the engine and grabbed the radio mic. "L.A., Engine Sixty-one. Respond a truck company with the jaws."

"Copy, Engine Sixty-one. Truck company for extrication," repeated the dispatcher.

Responding up Main Street to Brookfield Boulevard was the fastest route to the freeway. On the overpass ahead, Cole saw traffic stalled and cars lined down the on-ramp. Slowly, the paramedic squad and engine edged their way along the right shoulder of the freeway. Up ahead, on the center divider, a dull black mini-truck was embedded in the rear of a Caltrans dump truck.

Commuters slowed to gawk at the accident, further clogging traffic. Cole stood on the siren floor button, making it scream. Barker honked the air horn, causing traffic to stop completely. Slowly, the cars crept forward, creating a break, allowing just enough room for the paramedic truck and fire engine to cut across the freeway lanes. Barker positioned the fire engine in the number one lane, forcing traffic to divert around the accident, and allowing more room for the firemen to work.

Stepping from the fire engine, Cole was staggered by the devastation of the accident. The steel rear gate of the dump truck protruded through the windshield of the mini-truck. The parked dump truck had been pushed 20 feet forward. There were no skid marks from the little truck. The impact was so violent that the frame of the mini-truck was bent upward, suspending its rear wheels two feet off the ground. There was no movement inside the crushed cab.

The smell of gasoline was interlaced with the carnage. "Putnam, pull a protector line," shouted Cole. "Set the nozzle on spray and stand ready."

Putnam pulled a 50-foot hand line from the engine and stood ready to deluge the truck at the first hint of fire.

Mayor Jenkins shook Putnam's arm. "Spray the car! There's gasoline all around. Spray the car!"

"Do not shoot water!" yelled Cole.

Mayor Jenkins grabbed for the nozzle. Putnam pulled it away. The two started wrestling with the hose. Cole pulled Mayor Jenkins back. "Keep out of this, or I'll have you removed from the scene," said Cole, pushing Jenkins toward the engine.

Cole ran up to the accident site, weaving through a debris field strewn with broken glass, fragmented truck parts, baby clothes, knit booties and stuffed animals.

The two paramedics attempted to reach the victim squeezed tight between the steering wheel and the accordioned truck bed. The woman's tussled brown hair sparkled with slivers of glass, her face glistened sanguine with blood. Her right cheek was avulsed, revealing a mound of yellow fat tissue.

"Oh, my God! She's dead!" exclaimed the mayor.

Val took her wrist and watched her chest for rise. "She has a pulse," he said, surprised. "It's weak and thready." Valdez opened the oxygen tank and pressed the mask to her mangled face. He pushed the regulator button, forcing pure oxygen into the woman's lungs. With each pressured ventilation, air and blood spattered from her torn cheek. Valdez sealed her cheek with his gloved hand, reducing the air loss to a bloody mist blowing between his fingers. Her eyes were dull, dry and fishlike, and coated with powdered glass.

"Wash that glass off her face," demanded the mayor, leaning into the truck.

"Haven't got the time," said Cole. "That's not what'll kill her."

"Then do something!" demanded the mayor. He pushed past Cole and started barking orders at the paramedics.

Cole pulled Mayor Jenkins back by the shoulder. "I'll handle this. You back up, or I *will* have you arrested."

Mayor Jenkins begrudgingly stepped back but held his angry glare on Captain Walker.

"She's broken up bad," said Razmunson. "She's got a pulse, but that's about it."

Squeezing through the contorted passenger side window, Razmunson assessed the woman's condition, palpating her body from the legs up. The clutch pedal and brake pedal ensnared her severely angulated ankles. She was unresponsive with two broken femurs, two compound leg fractures, multiple broken ribs, a compound right arm and comminuted clavicle, severe facial trauma with raccoon eyes suggesting a probable skull fracture. Razmunson pressed his hand between her stomach and the bent steering wheel and palpated her abdomen. For an instant, her eyes blinked, then rolled upward exposing the white sclera.

"She's pregnant," Raz shouted in disappointment. He didn't like what he felt beneath his fingers. The woman's abdomen was asymmetrical, deeply concave in the center. Her crotch was soaked with blood, urine and embryonic fluid. The impact of the steering wheel had eliminated any hope for a viable fetus.

The paramedics were limited in their actions until she was removed from the vehicle. Spinal immobilization was difficult; CPR would be impossible. Razmunson set up a liter IV bag of normal saline, while Valdez pierced her anacubital with a large-bore needle. Captain Walker requested the hospital emergency room be placed on stand-by.

Truck 45 arrived, honking its way across the traffic lanes. With a few waves of his hand, Captain Decker had his firefighters assembling extrication equipment around the smashed truck.

"You need to cut the driver out," ordered Walker. "Start by cutting the doors off so the medics can work. Then remove the roof and the steering wheel."

Captain Decker relayed the order to his crew. The large truck firemen began dismantling the mini-truck with the Jaws of Life. The truck fireman jammed the tip of the hydraulic spreaders into the crack of the driver door and the powerful jaws tore the door from its hinges. Next, the second truck fireman cut the four roof posts and together they lifted off the roof. While the paramedics started a second IV, the truck firemen wrapped a chain around the dump truck chassis and another chain around the mini-truck's steering column. Then, he connected the loose chain ends to the opened jaws of the hydraulic spreaders. The truck fireman activated the powerful unit, closing the jaws, pulling the steering column through the dashboard and enlarging the paramedic's workspace.

Officer Parker reported to Captain Walker. He read from a driver's license. "Twenty-two-year-old female, Vanessa Arland. Witnesses said she was traveling eighty-plus, passing on the right shoulder and center median. She didn't see Caltrans doing road maintenance."

"Let's pull her out!" shouted Razmunson. He counted aloud to synchronize their movements. "One, two, three, lift."

With an even pull, the firemen slid the woman's flaccid body onto the backboard, but stopped short. The distal end of her protruding femur bone caught under the dashboard. Razmunson forcibly slid her exposed femur from underneath the dash. The bone shard squealed across the plastic dash like fingernails across a chalkboard. Mayor Jenkins recoiled from the crepitus and vomited.

Cole choked back his own nausea, not giving Mayor Jenkins the privilege of simultaneous hurls. He pursed his lips tight. Beads of sweat formed on his brow, while his stomach twisted as it considered reversing directions. Just in case Cole had to toss his lunch, he needed to find a discrete location. He turned abruptly and bumped into Captain Decker.

"She's broken up pretty bad," said Captain Decker, enjoying Cole's green hue. "Blood mixed with radiator fluid has a unique smell, doesn't it?"

Cole swallowed hard against the acidic burn in his throat.

Mayor Jenkins stepped between the two captains. Avoiding eye contact and redirecting his breath with his cupped hand, he said he had an important phone call to make and staggered off.

"First day on the job and already you're sucking butt," said Captain Decker, with a smirk and eyeing the mayor as he swaggered away. "Looks like your headquarters training is showing through."

Cole wanted to respond to Decker in a physical way but needed all his strength to keep his lunch down.

"Her ankle is stuck," said Razmunson, pulling at the brake pedal and tossing a bloody bunny slipper out of the truck.

"She's in full arrest," shouted Valdez, checking her carotid pulse. He forced more oxygen into her lungs, blowing blood and tissue through her cheek onto his partner's back.

The truck fireman forcibly wedged the hydraulic jaws between the floor pan and the brake pedal. The jaw tips spread easily, bending the pedal armature, releasing its grip of her foot. Her crushed ankle, held by strands of pasty white tendon, dangled freely as they slid her body across the backboard.

Putnam sidestepped along the rolling gurney and began CPR. His chest compressions met with an eerie lack of resistance; her torso quivered like gelatin. Valdez jumped into the back of the ambulance with Razmunson and raced away.

Putnam followed behind in the paramedic truck.

After the sirens had faded, Cole asked Barker, "Do you think she'll make it?"

"Not a chance," Barker answered, while loading the fire hose back on the engine. "Traumatic full arrests never survive. Today was just a drill...a multiple- company exercise in futility."

On their return to the station, Cole, Barker, Mayor Jenkins and Bleve rode in silence. Once in quarters, Mayor Jenkins, his mouth stinking of bile, walked backwards out the door as he spewed lame excuses of more pressing engagements. And, just as quickly as he came, he left.

"In like a lion and out like a lamb," said Barker.

About 30 minutes later, the paramedics returned to quarters. The bloodied equipment was washed and sanitized. The trauma box was restocked with bandages and IV bags. The backboard, sticky with blood, was replaced with a spare. Val still cooked dinner. The meal was served late.

The dinner conversation was slightly altered from its normal course of frivolity. For the first time in his seven-year career, Captain Cole Walker witnessed what happened after a tragic incident. He expected grief and remorse, but the table conversation was not emotionally charged.

The dinner chin didn't revolve around the tragedy of the accident, but of the procedures and protocol. The first topic was the patient's condition: compound fractures, internal bleeding, basilar skull fracture, cardiac tamponade and pregnant. No one, except Cole, was bothered by the dinner conversation's subject matter. On one hand, the firemen spoke of placenta abruptio, hypovolemia, crepitation and avulsions, while on the other handheld buttered bread and ate their peas and carrots.

Once the woman's condition was well established, the conversation shifted to their rescue tactics. Walker had made a good decision to not spray the truck with water, risking a static electric spark or spreading gasoline across the flowing water. Should the truck's roof have been removed before the doors? Should they have attempted to separate the two trucks to facilitate better access, but risk jarring the patient? Surprisingly, there was no grieving, no weeping, no tears, no sorrow, no remorse, no fasting—just cold facts tossed about the dinner table like they were discussing politics or their last vacation trip.

Engineer Barker summed up the incident and closed the conversation with his personal observation. "Looking on the bright side, an idiot driver gene was removed from the gene pool."

At first, Cole thought the firemen were morbid and insensitive for canvassing their observations so freely, but eventually, he realized that it all made sense. That's how firemen had to be. Firemen could not mourn their patients. They could not afford to fast or lament. They needed to remain detached from the tragedy or they would be consumed by emotion, becoming victims themselves. Cole thanked the firemen for their gallant efforts.

"Thanks for saving another organ donor," scoffed Barker.

Once the firemen were purged of the incident, the dinner conversation returned to normal topics and the crew dined as if nothing catastrophic had ever happened.

Following dinner, the firemen played four hands of cards. The last remaining player defaulted to washing the dishes. No pulling rank in the tank, was the only rule, effectively voiding Cole's only trump card.

Cole was ill-prepared for the firehouse card games, like West Hollywood where kings were queens and queens were kings; No Bowel Howell, a card game named after retired Chief Howell, whose estranged wife shot him four times in the abdomen, disintegrating his intestine. The card draw - take two, twice and the red queen kills your hand. It was obvious to Cole that he was going to wash dishes for a long time.

After the humiliation of washing dishes, the otherwise uneventful evening grew late. In the dayroom, the crew was enthralled with the finale of a pay-per-view movie. In the office, Cole read department training manuals.

Ding…

Cole calmly replaced his bookmark and stood to hear the alarm.

Ddddrrrrinnnngggg…

Engine Sixty-one, Engine Forty-five, Engine Thirty-one, Truck Forty-five, Squad Sixty-one. Manual pull alarm. Pioneer House. 11441 Pioneer Lane. Twenty-two twenty."

Cole ran to the fire engine. Bleve was already on the cowling barking and waging his stump of a tail. Cole kicked off his work boots. He stepped into his turnout boots and pulled up his turnout pants, taking a moment to check for panties. He was the first one dressed. In fact, he was the only one suited- up. Rus told Cole that the Pioneer House was always a false alarm and if he ever caught the bastard pulling the alarm, he would personally change the "fire response" to an "assault call".

Following Rus's advice, Cole nervously cancelled the other responding units. He notified the dispatcher that Engine 61 and Squad 61 would investigate the alarm. Barker fired up the engine and they headed for the Pioneer House.

Sixty seconds earlier at the Pioneer House, a 15-year-old boy darted across the street and dove into the bushes. The thick junipers were perfect for hiding and shook from the covey of anxious adolescent boys. "Shhhh!" came from the bushes. The *Third Alarm Dash* was the boys' favorite game.

Following the activation of the fire alarm, the boys hid in the juniper hedge of the Lutheran church and waited for the fire engine to arrive. After the fire engine's departure, the boys would laugh and recount their daring deed, then have a smoke.

The Pioneer House was a large log building that homestead farmers had erected as a barn. It was eventually converted to a community center complete with electricity and a manual pull fire alarm. When Engine 61 rolled up, the alarm bell was ringing.

The firemen got off their respective units and proceeded with their practiced routine. The paramedics peered inside the windows of the Pioneer House. Barker pushed back the pull box's activation handle. The deafening bell fell silent. Putnam replaced the keeper pin and the small pane of glass. Cole reported a false alarm and reset to dispatch. Cole paused for a moment, looked up and down the darkened street, then across to the junipers surrounding the church. No one was in sight. Satisfied, he climbed back into the engine cab. Barker shifted the engine into gear and reconfirmed that he would, "strangle the bastard, if given the chance."

After returning to the firehouse, Cole retreated to the office where he noted the day's activities in the station journal. He added superfluous details to his report, wiling away the time, and waiting for the crew to go to bed.

"Good night," said Putnam, sticking his bulbous face around the partially opened office door.

"Good night, Terry," said Cole. "Anybody else up?"

"Nope. We missed the end of the movie, and it's too late to start another, so we're all hitting the hay."

"Pleasant dreams," said Cole, offering a quick smile and returned his attention to the journal. When he heard the office door close, he shut the journal and reviewed his plan. He checked the wall clock and gave the firemen one hour to reach their REM sleep. He softly walked to the tool room and collected his materials. "Time to tighten the reins," he said, then crept upstairs.

Cole removed his boots before entering the dorm. He slowly closed the dormitory door and allowed his eyes to adjust to the faint illumination cast from the dispatch radio. Standing motionless in the darkness, he studied the room as his heart drummed in his chest.

Sticking to his plan, he would do the closest bunk first, Barker's bed. Then Raz. Then Val but skip Putnam. Cole figured Putnam was not involved in the coffee mug/boot polish incident, so he would be spared retaliation.

Slowly, Cole pulled a latex glove filled with powdered graphite from his jacket pocket. He carefully untied the pre-cut finger and poured a bead of graphite onto Barker's pillow around his head. As he poured, the bead of graphite slid down the pillow creases and disappeared behind Barker's neck. He moved to Razmunson.

Jerry Razmunson slept on his side. Cole poured the graphite directly into the cleft between the back of his head and the pillow, then he poured another mound directly in front of Raz's nose. He moved to Val.

Valdez also slept on his side. Approaching from his back, Cole poured a trail of graphite starting at the nape of his neck, encircling the top of his puffy hair, then around his face and ending under his chin.

Cole retied the knot in the glove and slipped it back into his pocket. He skulked past Putnam on his way to the fire pole. Putnam's snore concealed Cole's shuffling. Slipping up to the pole hole, his toes feeling for the edge, he reached out and grabbed the cool brass pole. He took a final glance across the dormitory, then back at Putnam who was sleeping clean and innocent. Cole pushed away from the pole and stood in the darkness, considering the ramifications.

Putnam was left untouched, different from the others. Although Cole didn't believe Putnam deserved his revenge, he did not deserve the label of *teacher's pet,* either. It may be worse for Putnam to be labeled as the captain's conspirator than to be graphited. So, Cole returned to Putnam's bunk.

Putnam's mouth was open and broadcasting a rhythmic snore. His eyes were cracked open, showing his glistening cornea through his eyelashes. Cole encircled Putnam's pumpkin-sized head with a dark powder line. The slippery graphite instantly flowed down the pillow into the deep valley created by Putnam's enormous cranium. Cole returned to the pole hole and slid from sight.

In the downstairs bathroom, Cole washed his hands. What little graphite got on his hands clung stubbornly and caused the soap to lather into gray bubbles for several washings. Cole took a long look at himself in the mirror. He was committed now. From this point on, there was no turning back.

Returning to the dorm, this time in his shorts and tee shirt, he was just as quiet as his first visit. He wanted to give the graphite as much time as possible to migrate and cover every area possible. Anxious, yet justified in his actions, he lowered his head to his pillow. His mind swam with the activities of the day: The Mayor; the traffic accident; his latest endeavor—revenge. Staring into the dark ceiling, he waited for sleep and prayed for a bell.

Rocked in the rhythm of Putnam's snore, Cole eventually drifted off to sleep. His turbulent slumber was marked by reenactments of the day's events. He returned to the traffic accident. This time, the young woman was reaching for him with bloodied fingers and pleading for help. The crew stood immobile by their equipment awaiting orders, but Cole couldn't speak. He went to help the woman but was held back by Mayor Jenkins who bellowed a hideous laugh. Cole struggled to free himself from the mayor's bony clutch as the woman begged louder and louder for help. Then suddenly, he and Mayor Jenkins were facing each other in a ring of smashed cars. The mayor stood in a bare-knuckled boxing stance. Cole's fists were tucked in bunny slippers. Cole stepped back, tripped over a large teddy bear and fell to the ground. Mayor Jenkins stood over him. Cole held up his bunny slippers in a feeble defense. Cocking his arm back, Mayor Jenkins swung down with a mighty blow.

Ding…

Cole shot upright in bed, his forehead moist with perspiration. He looked around, confused by the darkness

Ddddrrrriiiinnnngggg!

He recognized the radio dial as the dispatcher's voice brought him back to the fire station.

"Engine Sixty-one, Engine Forty-five, Engine Thirty-one, Truck Forty-five, Squad Sixty-one. Manual pull alarm. Pioneer House. 11441 Pioneer Lane. Two forty-two."

While pulling up his turnout pants, Cole studied the dormitory. He quickly checked his waist for panties, then stomped up and down the dormitory aisle. "Hurry up, ladies. Let's get moving!" he whooped, clapping his hands, hoping to add some confusion to their lethargy. "This could be the real thing."

Sitting in the engine cab, Cole grabbed the radio microphone and cancelled the other responding units. He told the dispatcher that Engine and Squad 61 would investigate the alarm.

"Someday, I'm gonna strangle that bastard," vented Barker, spinning the large steering wheel.

"Do I detect a little *friction?*" quipped Cole, smirking at his private joke. The engineer flashed him an annoyed scowl from a face glistening in the metallic gray lubricant. Barker looked like a peeved Tin Man.

They arrived at the Pioneer House to the same scenario. The pull box was activated, and the alarm bell was ringing. Cole pushed the handle closed, slid the keeper pin in place, replaced the small glass pane and turned to admire his crew. Valdez was studying himself in the engine's side mirror and poking at his streaked hair. The rest of the firemen were clustered on the sidewalk, examining each other, their faces shimmering gunmetal gray.

"Whoa," said Cole, staggering back, "you guys need your beauty sleep." He held his eyes on the firemen as he spoke into his portable radio to report the false alarm, then he stepped closer. "Let's have a lineup."

"Right now?" asked Valdez.

"Right here," said Cole, tapping his boot on the curb. "That's an order."

The firemen shuffled into line. They put their heels on the curb, causing their toes to sink into the soft grass. They stood at attention, teetering uncomfortably on their toes. They reminded Cole of doomed men standing on the gallows.

"I enjoy a good prank as much as the next guy. And, since you are the next guy, how are you enjoying my prank? I think our time together will go much smoother if you curtail your shenanigans. If you insist on playing vindictive games, I suggest you stick to poker or Monopoly. Well, gentlemen, I think that about *covers* it. Let's head back to the station and *slip* back into bed. Dismissed."

Cole felt vindicated as he slid between his cool, clean bed sheets. He lay still, enjoying the music of running water and murmuring. He reviewed the day's events over and over in his mind. Not feeling sleepy, he remained vigilant.

CHAPTER 6

OFFICER PORKER

Midday on his fifth shift, Cole backed through the kitchen door and set his armload of groceries on the table. He tucked a dishtowel into the front of his belt and began preparing the evening meal. This was his first turn as firehouse cook. His cooking rotation was after Valdez, who was a hard act to follow. Cole would have preferred following Putnam who was still developing a palatable menu.

Cole had elected to cook a time-honored Walker family recipe. The recipe had also been the longest in his cooking repertoire—about one week. He was well versed at preparing the meal because he had made it every off- duty day for the past week. His mom helped him the first time. He followed her recipe exactly and it came out, quite frankly, the way Mom used to make it. Unfortunately, the station atmosphere was different than home.

At the firehouse, criticism was sport. Putnam's last cooking shift turned hostile and uncomfortably long after he served tuna sandwiches for lunch and tuna casserole for dinner. After a brutal lambasting at the dinner table, the hot water was shut off during Putnam's evening shower, then his bedding was filled with cornflakes.

Cole's primary obstacle was the kitchen cookware. Most of the station's pots were huge, big enough to boil whole cattle. The utensils were army surplus. Leaving nothing to chance, he brought his own implements from home. Luckily, the emphasis on the evening's dinner would have a twist. Attention would not be on the meal, but on the dinner guest—Officer Porker.

As summed up by Jerry Razmunson, "Officer Parker does not excel in anything except waistbands, in which he exceeds."

Of all the qualities he lacked, Officer Parker excelled at eating. All You Can Eat food establishments did not relish his patronage.

Instead, it was his patrol car in their parking lot that inspired restaurateurs, the ones who could afford it, to entice Officer Parker with complimentary foodstuffs. His patrol route was in close concert with the baking schedules of local restaurants. And after a grueling day of feasting off the local merchants, Officer Parker invited himself to dine with the firemen.

Early on, behind his back, the firemen referred to him as, "Officer Porker". Eventually, the slanderous name surfaced and replaced his real one. The firemen used Officer Porker routinely to his face, in official reports and in public. Officer Parker received the insults as a sign of acceptance. He came to dinner more often.

The innuendos failed to satisfy the crew's contempt. Officer Porker merely shrugged off the badgering, smirking between mouthfuls, stuffing himself with seconds and thirds and fourths. Cole cautioned the crew to handle the situation with prudence because Officer Porker signed off their CHP fix-it tickets and routinely brought donuts to the firehouse, though no box ever contained a full dozen.

Thursday nights were Officer Porker's favorite night to eat at the firehouse. He didn't have duty the next day, hence, it was a night heaven made for over-indulgence at the firemen's expense. On Thursday nights, most of the firemen left the table hungry. Thursday was declared popcorn night because of the inevitable hunger pangs that followed Porker's visit. But this Thursday would be different. This Thursday would be Officer Porker Night.

Cole double-checked his ingredients: bacon, shrimp, real butter, whipping cream, linguini, bread sticks and a variety of exotic seasonings. As he prepared the ingredients, Razmunson was busy in the sink, preparing his boot.

As Cole de-veined shrimp, Razmunson, standing in one sock, cleaned his boot. He scrubbed the waffle sole vigorously with a brush and dish soap. He picked up the boot and smelled the sole, then offered it to Cole. Keeping his hands over the cutting board, Cole leaned toward the boot and sniffed the sole.

"Seems okay to me," he said, nodding his approval.

"I think I'll scrub it one more time," said Razmunson, not convinced. He poured more soap on the boot and brought it again to lather.

Cole began frying the bacon and sautéing the shrimp. He was not tempted to nibble at the meal. He had had it seven days in a row. The last few times, he gave the finished meal to his neighbors. They eventually offered it to their neighbors.

The meal preparation was proceeding exactly as practiced. The utensils were the same. The measurements were exact. The timing was identical. Cole felt confident that he had replicated conditions exactly as at home.

Ding...

Cole froze in panic. It was a crucial time in the preparation. He shut off the flame to the bacon. The cast iron pan was too hot to handle. He left the half-cooked bacon sitting in the grease, the same with the shrimp, leaving it to gel in the melted butter. He yanked the towel from his waist and stomped to the engine.

Ddddrrrriiiinnnngggg!

"Engine Sixty-one, Squad Sixty-one. Poisoning. 778 Quartz Avenue. Space Eighty-four. Sixteen forty-two."

"They better be hurting because I'm starving," snarled Barker.

Cole had difficulty concentrating on the incident at hand. His mind was back in the kitchen. When Barker parked the fire engine in front of the dilapidated mobile home, he squeezed Cole's shoulder and gently shook him back to the present.

The firemen walked up to the darkened trailer. A woman met them in the dimly lit doorway. The trailer interior was somber with only one small lamp in the corner of the living room.

"What happened?" asked Cole. The woman was too shaken to speak.

Razmunson knelt at the man sitting on the couch, "What happened?"

The man remained silent, staring at the floor, only occasionally turning his eyes to the woman.

"Ma'am, you have to talk to us, or we can't help you," said Cole. He put his hand on the woman's quaking shoulder and tried to calm her. "Why did you call us?"

The woman took a deep breath and whispered, "We had an argument and he…" She began to sob.

"What?" interjected Valdez, obviously perturbed.

The woman looked at her husband who glared back defiantly. "We got in an argument, and he drank some bleach." She collapsed in a chair and held her trembling hand over her mouth.

"You drank bleach?" repeated Razmunson, turning to the man. The man nodded and hung his head.

"You drank… bleach?" Razmunson repeated in disbelief. "Regular laundry bleach?" he asked. His voice carried the hint of sarcasm that he hoped it would. "Regular washing bleach? Scented or unscented?" he mused.

Cole recognized the deterioration of the conversation and interjected, "How much bleach did you husband drink? Can we see the container?"

The woman sprang to her feet and scampered to the laundry room, speaking over her shoulder, "I don't know how much. I was in the other room." She retrieved the gallon bottle and handed it to Cole. There was little bleach left.

"How much was in here before he drank?"

"I don't know," she cried, wiping her eyes, spreading streaks of eyeliner across her cheeks. "I think it was half full or maybe a little less." She moved her gaze back to her husband, who remained motionless staring at the floor.

Cole turned his attention back to the man sitting on the couch. His head was slumped between his knees; his shoulders trembled with each breath.

Razmunson knelt, put his hand on the man's shoulder and asked, "Can you tell me how much bleach you drank?"

The man looked up. Though he was young, his sallow skin and sunken eyes made him look much older. He whined with an eerie shrill from his burned vocal cords and shrugged his shoulders.

"One swallow?" asked Valdez, butting in. "Two? Three?"

The man nodded.

"Three swallows," said Valdez, turning to Captain Walker. "He drank three swallows," he repeated almost jubilant as if he won a round of charades.

Putnam gravitated between the paramedics. "Why?" he interjected, unable to control his curiosity. "Why would you drink bleach?"

"We were in a fight," the woman answered. "He threatened to kill himself if I left. I didn't think he would do it. I went to get my purse. It couldn't have been more than a minute," she said pointing down the hall then at her watch. "When I came out, he was sitting there with the bottle of bleach in his hand. He started drinking it." She put her trembling hand to her mouth and turned away.

The urgency of the situation was dawning on Razmunson. He turned his attention back to the man and lightly squeezed his shoulder. Razmunson leaned forward and asked in a low, concerned tone, "Can you talk?"

The man shook his head.

Razmunson put his hand on the man's stomach just above his navel. "Does it hurt when I press?"

The man lurched forward and vomited blood. The ejection of blood shot straight out of his mouth striking Razmunson in the chest, then strafed across his knees. Bloody bile splattered everywhere. The fresh bloody emesis was almost fluorescent red. The nauseating stench of bile and chlorine filled the small room.

Razmunson's chest burned as the blood and bleach saturated his shirt, adhering it to his skin. He tasted warm speckles of bleach on his lips as he stared down at his torso glistening with blood. The man paused to inhale, then violently spewed more fluorescent red blood.

"Get me an IV!" shouted Razmunson, as he pulled the drug box out of the puddle of bloody emesis. "Val, set up the MAST suit."

"Barker, you and Putnam put him in shock position," said Cole. "Ma'am, get a bucket or something."

Putnam and Barker laid the man on the couch and raised his feet to the arm rest. Cole waved for the ambulance crew to hurry with the gurney. The woman scampered back with a plastic bucket.

"Set up an IV!" shouted Razmunson, slapping the IV bag and tubing into Putnam's thigh. Putnam stood paralyzed, fixated on the vomiting man. "Set up the IV," repeated Razmunson, hitting Putnam in the groin with the IV package. Putnam snapped out of his trance. He tore open the saline's outer bag, plugged in the IV line and bled out the air bubbles. Razmunson slapped the man's forearm until his veins bulged to the surface of his skin, then he slid in a 14-gauge needle. The needle gave a confirming pop as it entered the vein. A stream of dark blood flowed through the open catheter; a sign it was in the vein.

"IV's ready," said Putnam, handing the tubing to Razmunson who plugged it into the catheter, stopping the oozing blood.

"Open it up. Wide open," said Razmunson.

Putnam released the restricting clamp on the IV line, a wide stream of saline shot through the drip chamber. "Good flow," reported Putnam, snapping the drip chamber with his finger.

Razmunson grabbed the man's other arm and established a second IV line. He barely flinched as the man rolled to his side and released his third volley of blood across the bucket rim. The pale and ashen man didn't have the strength to lift his head. His unblinking, lifeless eyes rolled upward. Streams of blood and saliva slung from his lips and dribbled down his neck as he slipped into hypovolemic shock.

Walker ordered the ambulance attendants to put the man on the gurney and fasten the anti-shock suit around his legs and lower torso. Putnam pumped air into the MAST chambers creating an auto-transfusion, forcing the last remaining blood from the man's legs into his vital organs. Razmunson squeezed the IV bags as he climbed into the ambulance with his patient. Cole slammed the doors, sending the ambulance racing to the hospital.

Spent from the ordeal, the woman braced herself in the doorway. She sniffled and smeared mascara across her cheeks. "Will he be okay?"

"We're doing everything possible, ma'am," said Cole, wanting to say more. The woman slid down the doorjamb, crumpled on the stoop and cried.

Cole helped her into the house and sat with her until family and clergy arrived.

On the way back to the station, Cole couldn't get the vision of the man out of his thoughts. Surely, he was going to die. He turned to Barker for consolation.

"Why would a man drink bleach?" pondered Cole.

"Because he's an idiot," snorted Barker, shooting a quick glance at Cole. "It's all part of the process of natural selection," he said, as he backed the fire engine into the apparatus room. He continued, speaking into the side mirrors, "It's nature's way of weeding out the morons."

Cole considered Barker's explanation without further comment. He eased himself from the cab and went upstairs to wash the gruesome memory from his hands and face.

When the paramedics returned, Razmunson was wearing hospital scrubs and fire boots. His hair was wet, and he reeked of disinfectant soap. He trotted upstairs to the locker room. Valdez hunted for Cole, eventually finding him in the bathroom scrubbing his face with cold water. Val's report came as no surprise. The man had been chemically burned from mouth to stomach. His internal bleeding was extensive and irreparable. There was nothing the ER doctors could do other than make him comfortable while he exsanguinated.

Cole braced himself over the sink and slowly shook his head. "Does everybody we touch die?" He dried his face, then brushed past Valdez without making eye contact. "I have a meal to prepare."

In the kitchen, Cole stared at the cold shrimp encased in solidified butter and the serpentine bacon locked in a sea of white grease. He found it difficult switching from one tragedy to another. With a sigh of resentment, he relit the stove and tried to salvage the meal.

Joining him on the other side of the kitchen and oblivious to Cole's catastrophe on the range, Razmunson whistled softly to himself, mixing his concoction. In a coffee mug, he stirred two spoonfuls of chunky peanut butter with some ground cinnamon, thyme and parsley. He stirred, examined his creation, and stirred some more. He folded in two pieces of raw elbow macaroni for sound effects. He scooped a wad of the dark brown mass on a spoon and held it toward Cole. "What do you think, Skipper?"

Cole twirled the spoon and inspected the goober. "Needs a bay leaf," he said.

"Right!" exclaimed Razmunson. "That would be the finishing touch." Razmunson rummaged through the cupboard, clinking through the seasonings until he found a bottle of bay leaves and gently swirled one into the mixture. He stepped over to Cole who was feebly poking at the slab of shrimp and solidified bacon. "Hey, Skipper, do you think I should bust up the bay leaf?"

"Bust it up," said Cole, not looking up from the pans.

"How does this smell?" asked Razmunson as he opened a small jar and raised it to Cole's nose.

"Whoa!" exclaimed Cole, dropping his spoon and fanning his nose. "What is that?"

"Catfish bait," Razmunson said proudly.

"It smells like crap," said Cole, pushing the jar back at Razmunson.

"It's supposed to," said Razmunson as he carefully resealed the jar.

Cole propped open the kitchen door to air out the room, while Razmunson taped the catfish bait jar under the kitchen table.

While the bacon and shrimp warmed, releasing themselves from their coagulation, Cole prepared the salad and pulled the pasta from the pot of boiling water. Eventually, he regained his composure and continued his recipe using the resuscitated ingredients.

Normally, Cole would have washed the starch from the pasta with hot tap water, but because of time constraints he skipped that step and poured the chopped bacon, shrimp and whipping cream into a bowl. He tossed the ingredients with the wooden pasta spoons he brought from home. The shrimp was spongy, and the bacon was overcooked but, all in all, the meal didn't look bad. Cole told Razmunson to set the table because dinner was ready.

Ding...

The alarm seized Cole with anger. "Not now!" he shouted at the alarm bell.

Ddddrrrriiiinnnnggggg!

"Engine Sixty-one, Engine Forty-five, Engine Thirty-one, Truck Forty- five, Squad Sixty-one. Manual pull alarm. Pioneer House. 11441 Pioneer Drive. Eighteen fourteen."

Jerking the towel form his waistband, Cole marched to the engine. He cancelled the other units. Engine and Squad 61 would respond alone to the Pioneer House.

"If I ever catch the guy, I'm gonna kick his butt," fumed Cole. While responding to the Pioneer House, Cole could only think of his ruined meal and homicide.

On scene at the Pioneer House, the firemen quickly looked in the windows, Barker replaced the broken alarm glass and Cole radioed his false alarm report. Then they headed back to the firehouse.

After hitting every red light on the way back, they eventually pulled into the station to find Officer Porker waiting. He was circling the table like a shark stalking his prey.

Bing! Cole rapped the alarm bell with a serving spoon. The famished firemen gravitated to their usual seats and directed Officer Porker to sit directly across from Razmunson.

As the firemen and Officer Porker slid into their seats, Cole placed a large platter on the table. He tried to pour the pasta onto the platter, but sticky with starch it clung to the bowl. He gently shook the bowl with no result. He had to pound on the pasta bowl with his fist until the wad of linguini fell onto the platter with a plop. The mound of pasta retained the shape of the bowl.

Cole poked at the pasta glob like it was a wounded animal. He threw the pasta spoons into the sink, then rifled through the cutlery drawer for a carving knife. He sliced the pasta mound and served it like pieces of birthday cake.

Cole's pasta surprise started Officer Porker salivating as he moistened his lips with his tongue. As Cole dropped a slab of pasta on Officer Porker's plate, he did a horrible impersonation of a French waiter, "Tonight, we are having shrimp scampi and bacon primavera vis a tossed romaine salad. Followed vis hot apple pie… a la mode, of course."

Cole then raised his water glass for a toast. "Over your lips and under your snout, we hope what goes in doesn't come out."

Officer Porker sat comfortably at the center of the table. He preferred center seating because he could reach all of the side dishes without having them passed. Plus, he could shortstop the condiments as they were passed from either end of the table.

Unbeknownst to Stanley Parker, another benefit of his center seating was it presented him with a ringside seat to the evening's entertainment. Sitting directly across the table was Jerry Razmunson who, just before dinner, smeared the odd concoction of chunky peanut butter, thyme and raw macaroni into the cleft of his boot heel. Together, the firemen and their guest of honor laughed and dined on Officer Porker's special night.

Before Razmunson began his performance, Officer Porker had to be placed in the proper mood. Barker started with his grotesque stories; detailed accounts of gory incidents like the guy who cut his own throat with a power saw and the girl who fell into an industrial corn vat and was pressed into tortilla chips. Seemingly unaffected, Officer Porker continued to eat, though his forehead was developing beads of sweat.

Valdez interjected his story of a man who dove headfirst into a commercial woodchipper. He compared the aftermath of the splattered entrails to the pasta sauce. Officer Porker slowed his eating and finished his last bite with an uncomfortable, audible gulp.

Cole joined in the dinner conversation, commenting on the cable history show he saw last night about Egyptians who mummified bodies by pulling their brains out through their nose.

"I knew a lady who once ate a skunk," said Putnam, eager to join in. The table fell silent. All eyes were on the rookie. "It was roadkill," he said trying to improve his statement. "With maggots," he added, then turned his attention back to his slice of pasta.

Razmunson unscrewed the lid from the catfish bait. The issuing stench flowed from under the table, filling the void between Valdez and Officer Porker. Val pinched his nose and crushed his thick eyebrows as he stared suspiciously at Officer Porker.

"It wasn't me," pleaded Porker.

Valdez stood for a moment, raising his nose to the air. "Smells like dog shit," he said. He sniffed his way around the table, eventually stopping at Razmunson. "It's coming from you. What's on your shoe?"

Razmunson lifted his foot and laid his boot across his knee.

"Not that one," scoffed Valdez. "The other one."

Razmunson raised his other boot and placed it on the table, so the Vibram sole was in full view of Officer Porker. The chunky peanut butter was caked with sprigs of parsley and basil. "Well, I'll be…" said Razmunson. "Bleve, did you do this? I must have stepped in it after our last call."

"Get your boot off the table," ordered Captain Walker.

"It's only dog crap, Skipper. It's not going to kill anybody," said Razmunson, picking out the bay leaves and inadvertently licked his fingers. Officer Porker edged back from the table.

"What?" asked Razmunson, looking around the table. "What? There's nothing scary about this stuff. It's natural. They say a dog's mouth is cleaner than a human's mouth. This just came from the other end." Razmunson scanned the table for support. "And besides, it's from Bleve. He's not diseased."

"If it's not so bad, then I dare you to eat it," challenged Valdez.

"I don't do dares," returned Razmunson.

"I'll give you twenty bucks, if you eat it," said Valdez.

"Chicken feed."

"A hundred dollars," said Barker, slapping his wallet on the table. "I'll give you a hundred dollars if you eat what's on your boot."

Officer Porker stopped chewing. He stared at Razmunson, then at Barker, then at Razmunson, like he was watching a tennis match. Porker's unchewed food began to drool from the corners of his gaping mouth.

Razmunson thought for a moment. "Hundred bucks, right?"

Barker nodded. "One hundred dollars."

"You're all witnesses," said Razmunson as he swung his finger at everyone at the table. "Stan, hand me a breadstick," he said, pointing to Officer Porker who just stared. Razmunson snapped his fingers. "Stanley... a breadstick, please."

Officer Porker remained dumbstruck. Barker handed Razmunson a breadstick. Razmunson bit the end of the breadstick and chewed it pensively, staring at the sole of his boot. Cautiously, he dug the breadstick into the gooey, brown wad then held it in front of his face, studying it. He confirmed with Barker the hundred-dollar bet, then plunged the breadstick into his mouth. He slowly rolled the breadstick around, smearing the brown concoction across his tongue. He then jutted his tongue at Officer Porker.

Officer Porker was frozen. The contents on his own mouth exceeded the plain of his dropped jaw, causing some partially chewed food to fall.

Razmunson used the breadstick to position the elbow macaroni between his rear molars. He bit down, crunching aloud the dry macaroni, slightly opening his mouth for proper noise projection. He contemplated, then jutted upright in his chair. "Hey, who's been feeding the dog chicken bones?"

Razmunson had planned to guess the brand of dog food Bleve had for dinner but was interrupted when Officer Porker vomited across the table. Porker's projectile interjection caught everyone by surprise, including Officer Porker who had no time to cover his mouth or direct his aim. His second volley streamed between his fingers clasped tight across his mouth. He hurled in the bathroom for the next 20 minutes.

Clean up went relatively easy, as firemen were known for preplanning disasters. The four corners of the tablecloth were collected, and the entire mess was carried outside and hosed off. By the time Officer Porker swaggered back from the bathroom, the kitchen was back to normal. He found the firemen dining on apple pie.

Razmunson was seated at the table, digging the remainder of the peanut butter from the cleft of his heel. As he licked the last of the gob from the breadstick, he said, "This isn't bad. You guys ought to try it."

Officer Porker raced back to the bathroom.

The firemen broke formation and quietly headed to their units. Cole was already in the cab when Rus climbed aboard. "Well, that went *smooth*. Rus, is it me, or are you turning *gray*?" Cole's smirk widened into a broad grin, then into laughter, admiring his handiwork and driving his point home.

After their return to quarters, Cole listened to the running water in the upstairs bathroom. He stuck his head in the bathroom doorway on his way to the dorm. Val, Raz and Putnam stood in front of the sinks, each covered with beards of gray lather. Barker stood naked in front of the shower. Long strands of hair hung from the left side of his head. He flipped his comb-over mane onto top of his bald head with a slap.

"You guys look *squeaky* clean. And don't forget to wash behind your ears," said Cole, holding a wide smile. "Sweet dreams, gentlemen," he said as he closed the bathroom door.

CHAPTER 7

THE COOKIE GIRL

Over the past several months, Captain Cole Walker had a chance to grow accustomed to the firehouse routine. The previous month, in particular, had no major fires, nor any dramatic rescues. The incidental band-aid calls and minor fender-benders faded one into the next. The only memorable incidents were the 350-pound naked man who decided to collapse against his tiny bathroom door and the lady who sprained her ankle when the shoestring she tried to hang herself with broke.

The C-shift began as monotonously as the previous dozen shifts. At ten o'clock, after exercising, the firemen assembled in the apparatus room for another drill, their third rope drill that week. Cole arrived last and found Barker teaching the firemen how to tie a noose. The front doorbell rang, and all the firemen dropped their nooses and clamored to the office.

The firemen pressed tight behind the captain as he opened the door. Their five curious faces broadened into five bright smiles as the door swung wide to reveal Evelynn Dewitt and a plate laden with cookies. Though she wore her predictably drab long dress and sandals, she was a welcome distraction to the tedium of their day.

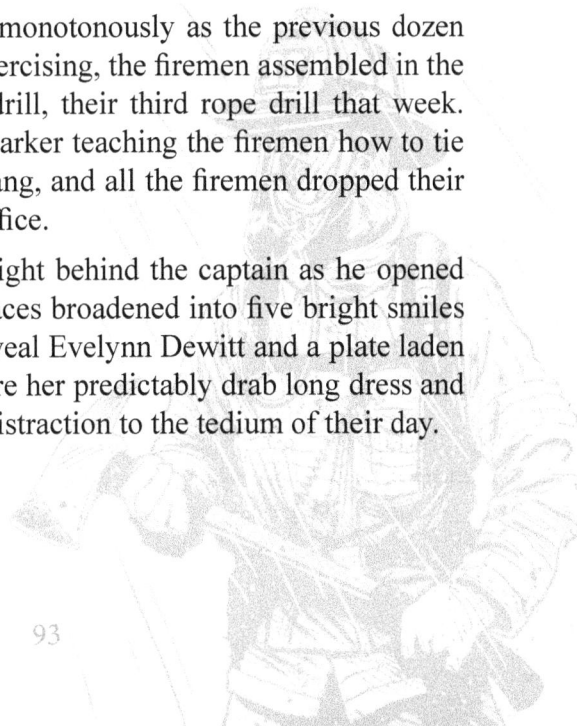

Ever since the station's open house, Evelynn's periodic visits, and cookies, became less of a surprise and more of a ritual. And although she didn't drop by as frequently as Officer Porker, she was much more welcome. The firemen dubbed her the *Cookie Girl*.

The C-shift was Evelynn's adopted crew. She enjoyed the firemen with innocent curiosity. She relished their wit and unique form of bantering. Evelynn felt as if she was peering behind the forbidden walls of the boys' gym or was an innocuous bystander in the corner of a private men's club. She did not take the privilege lightly. She only visited the station between two and four o'clock, never to disturb the firemen during lunch or dinner hours. She always left before dinner, careful not to intrude into their circle. At times, she felt as if she was conducting a study of primates—observing from a distance, holding still, inching closer as she cautiously approached the group. She watched them watch her and did nothing to startle them or betray their trust. As she was allowed closer, she rewarded them with cookies.

She was pleased at how deep she had infiltrated the firemen's lair. She made mental notes: *The more cookies I bring, the longer I can stay. Cakes are useful, too.* She mainly made chocolate cake because it was Terry Putnam's favorite, though he never said so. She could tell by the way he ate it, slower than the rest of his food. So, when she made a cake, it was definitely chocolate.

On the other hand, the firemen merely tolerated the Cookie Girl. She intruded in their space, but never stayed uncomfortably long. The cookies were a good excuse to sit and take a break from their tedious drills. The more cookies, the longer their break. At the moment, the firemen stood wedged in the tiny office staring at the mound of cookies in the hands of the comely Cookie Girl.

"Come in, Evelynn," said Cole, breaking the silence. He stepped back and bumped into the other firemen who were not able to move because Putnam blocked their egress. With an unobstructed view over the heads of the other firemen, Putnam stood transfixed, admiring the Cookie Girl. Their eyes met. He smiled. She grinned. He waved his fingers. She shrugged her shoulders and raised the plate of chocolate chip cookies. Their exchange of elementary salutations made the other crew members nauseous. Valdez was the first to say so.

"Stop your heavy breathing and back up, Putz."

The clot of firemen slowly dissolved, shuffling back through the narrow office door and reforming in the kitchen.

Evelynn's innocence made her a tolerable visitor. She was a new set of ears for the firemen's tattered stories and a new set of eyes that widened at each peril. She presented a ripe opportunity for embellished tales of rescue and narrow escapes that elongated with each recital—like the time Valdez held an angry mob at bay with a chainsaw until police arrived to arrest a neighborhood child molester; and the time Razmunson wrestled with an overcharged fire hose that slapped him around like an angry serpent until it tossed him onto the roof of the burning house.

Gullibly, she sat, wide-eyed and open-eared soaking in all that she heard.

Besides being an unwary visitor, Evelynn was also a new target. Anyone unaccustomed to the inner workings of the fire station, as Putnam could attest, was a likely candidate for raillery. Once the firemen learned to tolerate her, they showed their acceptance with ruthless teasing. Evelynn thrived on it.

The sacrosanct beliefs of the young woman were beyond reproach, except in the firehouse. The firemen challenged her beliefs and tested her resolve. She remained quiet when Valdez suggested using kittens instead of rats for laboratory experiments. She did not argue when Barker said marriage vows should include an exchange clause. She merely smiled when Captain Walker suggested wedding dresses be rented like tuxedos. Evelynn Dewitt endured the assaults, secretly vowing to never let the firemen make her angry, which infuriated the firemen. They promoted sport-hunting songbirds and bull fighting as national sports. Nothing irritated the Cookie Girl, until they attacked horses.

Barker's original suggestion was that horses should be raised strictly for meat and glue. There was no way Evelynn could sit idle and allow such a concept to go unchallenged. Her foremost dream was to have her own horse. The thought of cutting God's most beautiful creatures into steaks was sheer cannibalism and she made the mistake of saying so.

"Cannibalistic?" asked Barker, scratching his baldhead symbolically then smoothing his disturbed hair strands. "What makes eating horse meat cannibalistic?"

"A horse is your friend. You wouldn't eat your best friend, would you?"

"If it meant the difference between living or retaining a good friendship, thanks for the memories, but I'll be eating meat," returned Barker.

Formulating her retort, Evelynn stroked Bleve as he rested his head on her knee. She covered his ears and tried symbolism. "I suppose you could eat Bleve," she said, with a tone of displeasure only a scorned woman could affect.

"That's not the same," defended Barker. "Dogs chase away burglars and defend property. Raising horses for food is no more repulsive than raising cattle for the same purpose. You eat hamburgers, don't you?"

Evelynn sat frustrated. She crossed her arms, squeezing them tight against her chest.

Putnam feebly came to the Cookie Girl's aid. "I once saw a horse count," he said, looking back and forth between the two debaters. "The farmer had asked, 'What's three plus two?' and the horse tapped his hoof…"

"Horses have personality," Evelynn interrupted, squinting daggers at Barker. "You can keep your dumb opinion, however barbaric it is. I will never eat horse. Ever!"

She stuck her tongue at Barker, signaling the end of the conversation. Looking at her watch, she rose to her feet. "Well, I should be going." She said her good-byes around the table, making only curt eye contact with Barker.

"I'll walk you to your car," said Putnam, brushing cookie crumbs off his uniform. He picked up the empty cookie plate and shifted it to his left hand, allowing his right to sway freely next to Evelynn's available left hand. Putnam wondered what he would do next.

Ding…

He stood vexed in the kitchen doorway. *Of all the dumb timing.*

Ddddrrrriiiinnnngggg!

The dispatcher announced a manual pull alarm at the Pioneer House.

Cole ran to the engine to cancel the incoming units. Barker said something about being in the mood to kick somebody's butt as he slammed his coffee cup on the table. Putnam thought he would also kill the prankster if he ever got his hands on him. Giving the Cookie Girl a glance, then turning his eyes to his feet, he shuffled toward the fire engine.

He rushed back to the kitchen door, "Wanna wait a minute? We'll be right back." Evelynn said she would wait and took a seat at the table.

As usual, the call was a false alarm. The firemen quickly reset the alarm and returned to the station.

Evelynn stood from the kitchen table when she heard the bay doors open and the rhythmic beep of the engine's backup alarm. She sat in her chair, not wanting to appear anxious. Stretching to look and watching through the window of the kitchen door, she saw Terry Putnam take off his big turnout coat, then roll down his turnout pants over his large rubber boots. Briefly breaking her line of sight, she saw Razmunson dart upstairs. She eased back into her seat when Valdez pulled the door open.

"Great. I was hoping you were still here," said Valdez, with a broad, car-salesman smile. Putnam stepped through the kitchen door. Turning to Putnam and then back to the Cookie Girl, Valdez made an offer. "I have a few pictures left in my camera. Let me take a couple pictures of you two by the engine so I can have the film developed."

"Well, I don't know…" said Evelynn, stroking her lifeless hair.

"Come on. You look fine," said Valdez. He turned to Putnam. "Tell her she looks fine."

"You look beautiful," said Putnam, stepping back into the apparatus room. He held the door open for her. They all shuffled awkwardly back to the fire engine.

Valdez positioned the couple by the captain's door, centering the *E61* between them. Evelynn checked her reflection in the side window. She parted her straight, drab strands of hair like theater curtains, hooking them back over her protruding ears before she turned to the camera.

"No. That isn't quite right," said Valdez, stepping back and looking around. "How about standing in front of the engine?"

Chuckling at his fastidiousness, Putnam and Evelynn shuffled to the front bumper. Standing together, Putnam saw the opportunity to slip his arm around Evelynn.

"No. That's not quite right either," said Valdez, lowering the camera. He pulled the Cookie Girl's arm, positioning her by the left front corner of the engine, then pushed Putnam to the far-right corner. Both disappointed and both forcing smiles, they stood separated by the width of the fire engine.

Holding his camera to his eye, Valdez said, "Much better," stepping back onto the front apron. "Putz, take about two steps forward."

"Toward you? Away from the engine? Why?" questioned Putnam.

"Depth," replied Valdez. "Depth of field."

Putnam glanced sideways at Evelynn and, shrugging his huge shoulders, he obediently took two giant steps forward as if playing Mother-May-I. Valdez motioned Putnam forward by waving his hand, then halted him with his palm.

"Okay. Look at my hand." Val held his arm over his head and snapped his fingers. "One… Two… Three!" He dropped his arm and at the same instant Razmunson emptied his bucket.

Up on the station roof was Jerry Razmunson with a five-gallon bucket of water. He was waiting for his target—the top of Putnam's pumpkin sized head—to appear within his drop zone. On cue, he released his payload.

What the first of three high-speed photos caught was a wide smiling Putnam and a sheet of water only six inches off his left shoulder. Frame two was an explosion of water molecules as they atomized after impacting the top of the big man's head. The final frame was a somber firefighter, soaked in humiliation.

At first, they all laughed. Valdez, Razmunson and the Cookie Girl, but when Evelynn saw Terry's red cheeks and dismal expression, only Valdez and Razmunson continued to laugh. Evelynn rushed to Putnam's side. He gave her a short side glance from his hanging, dripping head, then returned his stare to his feet. "Well, that was a candid shot," she mused, offering a warm smile.

Laughing together and walking back into the station, Valdez and Razmunson swung their arms high overhead and slapped each other's palms. They disappeared into the kitchen, but their laughter still reached the front of the station.

A moment later, Valdez emerged through the kitchen door. "Hey, boot. Wipe down the engine. You splattered on it." He disappeared behind the door into another round of laughter.

Evelynn squeegeed Putnam's forehead with the palm of her hand. "Why do they pick on you so much?"

"Oh, it's all right. I'm the boot," he sighed. "I only have two months until my probation is over.

" Then will they stop?"

Putnam didn't answer.

CHAPTER 8

HORSEPLAY

For months, emergency responses for Station 61 were agonizingly sparse. Other than the four-year-old boy who got his privates stuck in his zipper, nothing of any consequence happened. By the end of the month, the boredom was tense. Even a response to the Pioneer House would have been a welcome break from the tedium.

Working wonted in the apparatus room, Rus Barker applied the sixth coat of wax on the fire engine. Razmunson reorganized the supply drawers on the paramedic truck. Cole retyped the labels on all the fire prevention inspection files. Busy work no longer subdued the neurotic crew. Horseplay surfaced in innocent larks, but small pranks grew exponentially with each following payback.

During the first week of horseplay, somebody filled the dishwasher with regular dish soap causing the kitchen to fill with suds. Cole found a note on his desk to *Call Mr. Lyon.* When Cole dialed the number and asked for Mr. Lyon, he was informed by the curator that he had called the L.A. Zoo.

Putnam opened his locker door, triggering a bucket of water to dump on his head. Razmunson's running shoes were nailed to the bottom of his locker. And as Valdez showered, somebody dumped a bag of flour on his head.

The next shift, high jinks began early. Barker discovered ground garlic had been injected into his toothpaste. At morning line-up, Cole's coffee mug handle had been coated with honey. During lunch, Putnam set his deluxe Gyro sandwich on his dinner plate that had been filled to the rim with water. The contents of the saltshaker dumped into Valdez's soup when the lid fell off.

Before dinner, Valdez taped a 50cc syringe under the table and ran enough IV tubing to reach the end of the table, just above Barker's seat. During dinner, every time Barker drank, Valdez shot water onto his crotch. Each time, Barker would wipe his lap then check his cup for leaks. He threw away two cups before he looked under the table and got sprayed in the face. Barker laughed along with the rest of the crew and continued his meal. He finished early, made a quiet exit and returned with a charged hose line. He blasted the food from the table and the firemen from their chairs.

With each passing day, the pranks increased in frequency and ingenuity. In the morning, Cole found his shirt sleeves sewn closed. A small, dead frog on a three-foot string was strung inside Barker's pants, then re-hung in his locker. When Barker felt something creepy moving around his thigh, he leaped around the locker room kicking off his pants, nearly cold cocking himself on the towel rack. A 35-gallon trash bag was suspended in Putnam's locker and filled with water through the door vent. Putnam opened the door and was washed into the hallway. Valdez was late to discover that his hair spray bottle was filled with ink. At midnight and midstream, Barker discovered the toilet bowl was wrapped with cellophane.

The same battle against boredom being fought at Station 61 was also being waged at Station 45. It was during this emergency arid cycle that Truck 45 covertly ventured into Brookfield. While Cole and his crew were exercising at the Brookfield city park, Decker and his crew paid an unannounced visit to Station 61.

When Cole and crew returned to quarters, they found Bleve cowering under a recliner chair. They repeatedly called to him, but he would not come out. With all five firemen coaxing from their knees, Bleve skulked from under the chair with his head hung low and his stubby tail tucked between his legs. A large *45* was stenciled on his flanks. Putnam gingerly carried Bleve to the mop sink for a bath.

"I'll ring that son of a bitch's neck," steamed Valdez.

The phone rang. Cole picked up the phone and immediately recognized the mayor's voice as it entered his spine. "Captain Walker…"

"I'm sorry, Mayor Jenkins, but this is not a good time," interrupted Cole.

"Sorry to take you away from your circle jerk, but I just wanted to thank you," said Mayor Jenkins with icy cheer.

"For what?" asked Cole.

"For giving me a campaign platform," sneered the mayor. "As you know, I'm running for re-election. I'm going to campaign *against* L.A. Fire Protection. Want to hear my slogan? *Fire L.A. Fire*. Not bad, huh? I think it has a nice ring to it."

"I think you're being a bit hasty, Mayor. You haven't seen what LA Fire can offer," offered Cole.

"I've seen enough, boy. And, with every screw-up you make, I'll get free publicity. Besides, I've already ordered posters and campaign buttons. Have a nice day." The phone line went dead.

Cole spent the rest of the shift agonizing over Mayor Jenkins' revelation. The crew spent the rest of the shift stewing over the Bleve caper. The station ambience was uncomfortably hostile. As the day dragged into night, the mood throughout the station soured.

Shortly after midnight, the alarm bell disrupted the serenity of their troubled slumber. The dispatcher announced a "choking victim". The paramedics were obviously faster dressing and departing. Struggling to awaken from his sleep, Cole discovered he had slept on his kinked arm. He struggled to pull up his turnout pants with his left hand, while his numb right arm swayed flaccid. Cole clutched his loose pants with his left hand and trotted to the fire engine.

Exactly four minutes later, the squad and engine rumbled to a stop in front of a two-story house. A blond woman with dark roots waved hysterically. Cole ordered Putnam into the house to assist the choking victim, while the balance of the crew followed with the first aid equipment. The blond woman with the dark roots led Putnam upstairs and into the master bedroom.

Panting to catch his own breath, Putnam found a woman in her early thirties sitting on the edge of the bed. Her head was in her hands. Putnam shook her shoulder.

"Can you speak?" he shouted.

The woman jumped back startled. "Oh! You scared me," she exclaimed, fanning her face with her hand.

"What's the problem?" asked Putnam, speaking between his labored breaths.

"I have a dry tongue. I can hardly speak," she replied and stuck out her tongue. "Touth it," she said, pinching her tongue between her fingers. Her eyes crossed as she tried to focus on her protruding tongue.

Valdez and Razmunson, obviously out of breath, ran into the bedroom. They set down their armloads of first aid equipment.

"Is she okay?" asked Cole, sticking his head between the two paramedics. He stared vexed at the woman, who sat oblivious to the firemen as she drummed on her tongue with her fingers.

"Her tongue is dry," said Putnam.

"What?" asked Cole, pushing his way between the paramedics. "Your tongue is dry?"

"Yeth," nodded the woman, cross-eyed, pinching her tongue. "Touth it," she said, leaning forward, pointing her tongue at the captain. "Ith's been dry for almotht three hourth."

"Our dispatcher said you were choking. Are you choking? Are you having difficulty breathing?" His voice rose in volume, revealing his irritation.

"No," said the woman. "But I can't get a good night thleep like thith," she said, pointing to her tongue, then to her watch.

Cole took another step toward the woman and shook his finger at her when Razmunson caught his arm.

"Hey, Skipper, we can handle this. You guys can go back to the station."

Cole glared at Razmunson, momentarily blinded by his anger before realizing that Razmunson was right. He was not in the proper mood to handle this situation diplomatically. Cole waved for the engine crew to follow him as he stormed from the bedroom.

Humored by the doltish woman, Razmunson knelt by her side and unleashed his medical assessment. "Ma'am, I have a series of questions that sound odd, but are necessary so I can evaluate you. Will you help me?"

The woman nodded and released her grip on her tongue.

"Have you had any stimulants this evening… coffee, tea, adult movies?"

"Uh-uh."

"Ma'am, does this tongue dryness come with aberrations like seeing spots or flying toasters?"

"Uh-uh," she resumed squeezing her tongue.

"Have you eaten anything unusual? Thai, Greek, or Norwegian foods?"

"Uh-uh."

"She had a lime Popsicle," blurted the blond woman with dark roots, jutting her finger in the air to dramatize her brilliant recollection.

Razmunson nodded seriously and repeated "Lime Popsicle" to Valdez who repeated, "Lime Popsicle" again and wrote it on his report form.

"Do you prepare your own food? Would anyone want to poison you?"

The woman scrutinized the blond woman with the dark roots before shaking her head. "Uh-uh."

"Think, ma'am! Are you sure?"

The woman looked suspiciously at her roommate and raised an eyebrow. "Uh-uh."

"Are you testifying at a federal investigation? Or party to any litigation against organized crime?"

The woman searched the ceiling for her answer. "Uh-uh."

"Have you taken any cold or sinus medicines lately?"

"Uh-uh."

"No medicines at all?"

"Just for alergieth."

"*Alergieth?*" repeated Razmunson

"Allergies," translated the blond woman with the dark roots. "You know, runny nose," she said, pointing to her own face. "She took a lot of these last night." She pulled a bottle from the dresser drawer and handed it to Razmunson.

"This has diphenhydramine in it," said Razmunson. "How many did you take?"

"Theveral."

"*Theveral?*"

"Yeth."

"You exceeded the recommended dosage. In medical terms, we call your ailment lizard tongue. The good news is that you're not going to die. Unfortunately, you're going to have to let the meds run their course."

"But how can I get thome thleep?"

"I suggest warm milk and a half hour of Lawrence Welk. Good night."

The paramedics left. Riding back to quarters, Valdez vented his anger about the woman and her dry tongue. "Now I won't be able to thleep tonight."

"Since we're both awake, want to make a little side trip?" asked Razmunson, batting his eyebrows. Without waiting for an answer, he squealed the paramedic truck into a hard right and headed toward Station 45.

On the way to the station, Razmunson described an old college prank to Valdez who thought it wasn't mean enough. He would prefer boiling Decker in oil. Razmunson shut off the truck lights and coasted to a stop around the corner from Station 45. Clutching some wood cribbing and a ring of keys, Razmunson looked up and down the quiet street before tiptoeing to the back wall. The two medics climbed the brick wall and surveyed the station's dark parking lot before jumping over. Using the departmental keys, Valdez unlocked the tool shed and they searched within the narrow beams of their penlights. Quietly, they carried the floor jack from the shed.

Next, they examined each of the parked vehicles. Moving from one to the next, Razmunson waved to Valdez. "This is it," he whispered, pointing to a silver Polaris. It had a USMC decal in the window and a bumper sticker that read, *if firemen were more like Marines, fires wouldn't dare start.*

Quietly rolling the jack underneath the differential of Captain Decker's car, they raised the rear wheels a quarter inch off the pavement and blocked the axle with the wood cribbing. They replaced the jack in the shed and locked the door. Razmunson pulled a toothpick from his jacket pocket and broke it in the shed lock. They shimmied over the back wall.

Starting the squad and driving away without headlights, Valdez and Razmunson slapped each other's palms and laughed with anticipation. They knew they wouldn't see their prank unfold, but they laughed and slapped palms again just thinking about the morning's event.

At precisely 0705 the next morning, just as they had anticipated, Captain Decker started his car and returned to the station for a cup of coffee, while his motor warmed up. Returning with his USMC commuter cup, he sat behind the wheel and slid the transmission into Reverse. The car did not move.

Outside, the tires spun mere centimeters off the ground. Razmunson and Valdez had figured Captain Decker would notice the tachometer's high RPM and speedometer showing 60 mph., then he would discover that his car was on blocks. He would be stranded, locked out of the tool shed, vulnerable and embarrassed in front of his crew. For the most part, that is what happened—with one exception.

When Captain Decker saw his high RPM, he did not notice the speedometer showing 60 mph. Nor did he notice his tires spinning. What he did notice was he was not moving. Instinctively, he pushed the automatic transmission lever into *Park*.

The ensuing noise of the transmission shifting from 60 milesper-hour to instant stop caused firemen to come outside running. Decker's crew witnessed the aftermath of destruction as pieces of transmission rained to the ground and roll in every direction of the compass.

At 0720, Captain Decker was on the phone to Captain Walker.

"Good morning. Fire Station Sixty-one. Captain Walker," said Cole, dripping wet from the shower. He pinched the receiver between his head and shoulder as he tightened the towel around his waist.

"Wal-ker!"

Cole dropped the receiver from his ear and the towel from his waist. While Cole fumbled with his towel, he heard Captain Decker screaming from the phone lying on the locker room floor. Decker cussed for over two minutes, maybe longer. Cole was not sure exactly how long Captain Decker cussed because he left the phone on the floor, while he left to put on his pants. When he returned, he could hear Captain Decker from the doorway.

"Hello? Walker! You better answer. Wal-ker!"

When Cole picked up the phone, Captain Decker began cussing again. Cole jerked the phone from his ear and held it at arm's length. He could not understand a single word, nor could he insert one. Cole lightly tapped the receiver on the wall, checked for cussing and tapped the receiver on the wall again until the cussing subsided.

"Good morning, Captain Decker. What can I help you with?" asked Cole, as he snapped the phone away from his ear knowing that the cussing would resume. He held the receiver in his outstretched arm to reduce Captain Decker's voice to an audible level.

Between the profanities, from what Cole could decipher something had happened to a transmission from Captain Decker.

"We didn't receive a transmission from you," Cole yelled toward the mouthpiece.

"Not a radio transmission, you idiot! A car transmission," shouted Captain Decker, which he further clarified with profanity. Cole laid the phone on the desk and called for a line-up.

The four C-shift firemen slowly and sleepily meandered into the kitchen.

Cole told the crew he just received a disturbing call from Captain Decker. From what he could extrapolate, practical jokes had reached an epidemic level at Station 45. Cole was worried the same thing could happen at their station. He wanted a truce. But the firemen resisted.

"We can't stop cold turkey."

"We've got to go out with a bang."

"Can we pick on somebody outside our station?"

"Like who?" asked Cole.

"Like Decker for screwing with Bleve."

"Captain Decker is already wound too tight. He is one click away from barricading himself in a clock tower. So, no," said Cole.

"How about Mayor Jenkins? He's a jerk. He deserves it."

"If we got caught, it would be the end of the annexation. So, no."

"Officer Porker?"

"Already done. No."

"How about the Lady's Auxiliary?"

"They'd all have heart attacks. No."

"How about the Cookie Girl?"

"What?" asked Cole, astonished. "Why would you want to pick on the Cookie Girl?"

"First of all, she won't die," said Razmunson. "And it's obvious that she likes to get teased."

"She's here just about every shift, so hell, let's initiate her," added Valdez. "Besides, Putnam's got the hots for her. A clever prank would be the icebreaker that would give them something to talk about. We would be doing Putz a favor."

"And she's a good sport," added Razmunson. "I bet she laughs harder than we do."

Cole turned to Putnam who was watching from behind the refrigerator door. "Terry, what do you think?"

"Well, I was looking for an opportunity to ask her out. Do you really think she'd laugh?" asked Putnam.

"Absolutely," said Valdez, unsure.

"Positively," said Razmunson, not sure either.

Putnam probed suspiciously, "What do you plan to do to her? Dump water on her or light her shoes on fire?"

"I have an idea," interrupted Barker. "But it'll take everybody's participation."

All eyes turned to Putnam. He didn't want to join their cabal, but he believed Evelynn was a good sport and maybe a clever ruse would be the romantic opening he was looking for. He conceded.

Barker laid out his plan. Putnam's participation was crucial. He had to invite the Cookie Girl over for dinner next shift. When Putnam called Evelynn that night, she didn't sound suspicious. Instead, she was ecstatic about the dinner invitation. Her perky voice and excitement made him feel sick. Putnam hung his head when he hung up the phone, setting the plan in motion.

Atlanta, Georgia - Dennis Upman picked up his primer gray Kenworth tractor and trailer from the repair shop. The repair bill was more than he expected. It always was. The price for the transmission overhaul, air brake repair, and 16 new tires sent him stumbling backwards. The amount maxed his credit card. He wouldn't be able to buy a new radio, much less a new rig, until he made some serious money.

He climbed into the cab and fired up the diesel engine, blowing black smoke from the twin exhaust stacks. He flipped his finger to the repairman and lunged from the repair yard. He needed to make one more stop before hauling his load west.

Putnam stared at the *Cats* theater tickets in his hand. He wanted to take Evelynn some place special but wasn't sure exactly what she'd like. He figured *Cats* was about cats and cats were animals and Evelynn liked animals, so Putnam hoped she'd like *Cats*. The performance was still several months away but the tickets were front row seats. And Evelynn deserved front row.

The headlights of Evelynn's car swept across Putnam as he paced in front of the firehouse. He slid the tickets into his pocket and opened her door.

Evelynn greeted him with a wide smile and a warm apple pie. Putnam held the warm pie in his large hands and savored the sweet aroma of cinnamon and apple. Focusing beyond the pie crust, Putnam noticed the black high heels strapped to long, slender, silky legs. He had to swallow twice to control his salivation. Evelynn was dressed up.

Instead of her predictable long, drab, shapeless Quaker dress, Evelynn was wrapped in a mid-thigh, tight charcoal-gray skirt with cocoa nylons, all perched on three-inch stiletto heels. She stood from the car and came eye-to- eye with Putnam. He fumbled with the pie, nearly dropping it. Evelynn was wearing makeup. Putnam studied her new face. Her rose petal cheeks and deep blue eyes were framed in shimmering curls of golden hair that rested atop her lightly freckled shoulders. Her beautiful teeth teased against her strawberry lips as she placed her hand inside his elbow. Then, Terrence Elliot Putnam escorted Evelynn Sophia Dewitt into the firehouse.

When she made her graceful entrance into the kitchen, Bleve was the next to greet her. Bleve wore a black bowtie and a black silk cape. The cape was an afterthought, mainly to cover up his stenciling. When Evelynn saw the dining table, her cheeks flushed, and she let out a little squeal. Clasping her hands tight between her breasts, she surveyed the arrangements. The table had been covered in fine white linen. Instead of the usual paper hand towels, there were real cloth napkins folded into fans, anointing the thick Buffalo China and giving grace to the dulled everyday utensils. The plastic water cups were replaced with crystal goblets. Two bottles of sparkling apple cider were chilling in silver ice buckets at each end of the table. And towering above the arrangements were two flickering candles in new silver holders that Putnam had bought special for the occasion.

The Cookie Girl slowly strolled around the table, carefully absorbing each detail. Bleve followed at her heels, acting as the official Maître d'. While Evelynn studied the setting, the firemen studied her. She looked up unexpectedly and caused everybody to spin back to his assigned duty.

Putnam lowered his head, fearing that she would see him blush. The Cookie Girl released another giggle. Her excitement was obvious. She was a great guest of honor.

"Sit...Sit," insisted Cole, sliding out a side chair, the one next to Putnam's. Evelynn stood in front of the chair, slipping her sweater from her shoulders. As Putnam gently draped the sweater across the back of her chair, the brush of his hand against her bare shoulder made his blush return. Once the Cookie Girl was seated, the firemen took their places around the table.

Barker gyrated around the kitchen, preparing this, stirring that. He spun around producing an appetizer of sautéed shrimp still sizzling in the skillet. The Cookie Girl let out another delightful squeal. Putnam served her plate. Razmunson poured the sparkling cider. Valdez helped Barker in the kitchen.

With impeccable timing to the consumption of the last shrimp, Valdez placed the entrees on the table—a pot of fresh buttered green beans, scalloped potatoes au gratin, creamed corn, and hot bread rolls. Barker pushed through the back door with his hip and laid down the main course: A silver platter of deep red steaks garnished with sautéed mushrooms sizzling hot off the barbecue. The Cookie Girl unleashed another squeal and clapped her hands in approval.

Captain Walker tapped his fork against his goblet, then raised it for a toast. Everyone hushed and followed his lead. "I propose a toast. To our good friend and cookie aficionado, Evelynn Dewitt. Thank you for putting up with us. Thank you for your friendship. And thank you for the cookies. To Evelynn."

"To Evelynn," echoed the crew and clanked Barker's fine crystal goblets, toasting their guest of honor.

The meal was better than Evelynn could have possibly dreamed. She took her time with each entrée and sampled everything. She loved the beans with bacon and onion. She wanted the recipe for the scalloped potatoes. The corn and fresh baked rolls were heavenly. The marinated steak was exquisite. She wanted to know the cut of meat.

"It's a shoulder cut... the most tender," smiled Barker. *"Nothing but the best,* I always say... because I believe you are what you eat." He raised his goblet and clinked crystal glasses with the Cookie Girl.

The dinner and evening were elegant and delightful. Evelynn cleaned her plate and had seconds, saying the whole time that she shouldn't. As her plate grew empty, the firemen made a point to pass more meat to her. Evelynn complimented Barker on his cooking talent and thanked him for taking the time to prepare such an exquisite meal.

"Oh, I don't mind getting saddled with the cooking," he said, with an unnatural grin. About the time Evelynn pushed away her plate, the firemen started phase two.

They began by picking apart Barker's meal; first, by berating the preparation, then the inferior quality of meat.

"Hay don't nag me," he snapped back. But Barker's angry response made all the firemen laugh, including Barker himself. Barker got up curtly from the table and softly whinnied. When Putnam brought Evelynn's plate to the sink, Valdez jumped on his back and throttled him with a wooden spoon as he trotted around the table.

"Yee haw!" said Valdez, as he spurred Putnam in the ribs. "Giddy up!"

"Stop horsing around," said Cole.

"There's nothing wrong with a little horseplay, Cap," said Razmunson, chewing noisily on a large raw carrot.

The Cookie Girl's expression turned from naïve amusement to wide eyed realization that she had eaten horse meat. Her stomach suddenly felt very heavy, then knotted, forcing a scowl to her face. She felt heat from the blood pumping to her cheeks. She could not speak. She didn't know where to start or what to say.

"You...you..." she stammered, pointing at each of the firemen with the piece of bread crust still in her hand. She stood and snatched the sweater off the back of her chair. Growing angrier, she repeated, "You...all of you..." She gave a special glower to Putnam, squinting her eyes and focusing her daggers, "And, you!" She swung her sweater over her shoulder, knocking over some goblets, and stomped from the kitchen.

The kitchen door slapped back and forth, then fell silent. The firemen stood in silence listening to the hollow sound of high heels stomping across the apparatus floor, then through the front office. Even knowing what would come next, they all jumped when the front door slammed. The kitchen erupted in laughter. Valdez and Razmunson gave each other a hand slap. The kitchen reverberated in laughter as everyone took turns reenacting the antics and conversation leading up to the Cookie Girl's enlightenment.

On the other side of the room, Putnam closed his fist around the *Cats* tickets. He spread open the window blinds and quietly watched the Cookie Girl drive away.

"Damn," said Razmunson.

"Pay up," Valdez demanded as he held out his hand.

Razmunson pulled out his wallet and slapped 20 dollars into Valdez's palm.

"What's that for?" asked Cole.

"I thought she'd hurl," said Razmunson, looking into his empty wallet.

"Hey," interrupted Barker. "Who wants apple pie?

CHAPTER 9

RECONCILIATION

Dennis Upman set the parking brake with a loud blast of air, then hopped down from his big rig. Standing and stretching, he surveyed the truck stop, diner and souvenir shop. Scanning the outer fringe of the complex, he saw the man he was looking for lurking beside the repair garage.

"Yo, dude!" said the man, waving the trucker into the shadows. The two men performed their ritual handshake, hand-slap and fist-rub. The trucker pulled a fold of money from his pocket and exchanged it for a small brown vial. He shook three of the small white tablets into his hand and studied the crudely formed, but potent, amphetamine.

"That should keep you truckin' for a while, dude," said the sinister man, his face still in the shadows.

Upman resealed the vial and put it into his jacket pocket. He raised the pills to his lips.

"Whoa!" cautioned the dealer. "That's powerful shit, man. That'll keep you high for two days. Just take one."

Upman considered the dealer's advice before dropping two of the pills into his shirt pocket. He threw back his head and swallowed the single tablet as the dealer slid back into the shadows.

Dennis Upman felt wonderful and invincible as he climbed back into his truck. He fumbled beneath the seat. He pulled out three books. Shuffling through the counterfeit travel journals, he picked one and slid the others into hiding. He released the brakes and pointed his truck westbound.

It had been several weeks since the Cookie Girl's last supper at the firehouse. Every shift, Putnam would leap to the front windows at the sound of a car door. It was not until the third week after the horse meat incident that Evelynn returned to the fire station. Putnam opened the door, expecting another Boy Scout selling popcorn or a lost motorist looking for directions, but instead he stared into the soft blue eyes of Evelynn Dewitt.

Putnam blurted out his greetings. "I'm sure glad to see you again," he said, frozen, blocking the doorway. Evelynn looked even better than Putnam remembered her. Her beautiful golden hair, slightly curled at the ends, was lustrous in the sunlight. She had softened her lip shade from strawberry to a pink peppermint, and her cheeks were radiant with natural tone. She waited patiently for him to overcome his surprise. When he finally realized he was staring, he stepped aside and invited Evelynn into the station.

As she strolled past, the delicate redolence of rose and orchids wisped across his cheeks like a cool caress. Her chiffon sundress fanned with each step, flashing her long, silky legs. She spun to face Putnam, catching him in mid-gawk. He turned quickly to close the door and hide his embarrassment.

Evelynn ignored his redness and apologized for staying away so long. She said the more she thought about their last dinner, the more she felt she had handled the situation poorly. "I'm sorry," she said, holding out a chocolate cake in one hand. Clutched against her breasts, she cradled a paper bag with two half gallons of milk.

"Come on back," stammered Putnam, struggling to control his enthusiasm.

Putnam yelled for the crew, his voice booming across the apparatus room.

Barker appeared from around the fire engine and wiping polish from his hands, he followed them into the kitchen. Putnam seated Evelynn, excused himself, stepped into the apparatus room again and yelled for the rest of the crew. Valdez and Razmunson came in shortly, followed by Captain Walker with a fire department magazine rolled up under his arm. For a seemingly endless period of time, they all sat at the table sharing an uncomfortable, awkward quiet. Evelynn broke the silence.

"Look," she started, "I'm sorry for getting so upset the other day. I think I handled the situation poorly. I hope there are no hard feelings."

"No. No, there are no hard feelings," insisted Putnam, quickly looking around the table for consensus.

"Actually, looking back, it was pretty funny," she said, with her familiar giggle. "Especially when Terry trotted around the table. I should have figured it out when Rus said, 'Don't nag me.' Duh," she added, playfully slapping her palm to her forehead. She pushed the devil's food cake to the center of the table. "Here's a little peace offering. I hope we can still be friends."

"We're still friends," returned Putnam, nodding heavily to everybody around the table. He spun to the cupboard and rummaged for plates. Evelynn cut the cake and gave everybody a big slice.

"Don't you want any?" asked Putnam, speaking around the wad of cake in his mouth.

"No, thanks," said Evelynn, sliding her slice of cake to Putnam with a wink. "I must watch my weight. I'm getting as big as a horse." She slapped her thigh. The room erupted in laughter as close friends enjoyed each other's company once again.

Though nobody said so, the firemen thought it was nice to have the Cookie Girl back in the firehouse. Their feelings were evidenced by their laughter. All the previous tension dissipated as they reminisced about the *horse dinner*. Everyone agreed that Evelynn was a pretty good sport. She split the remainder of the cake amongst the firemen so she could take her great aunt's antique plate back home. Putnam washed and dried it for her.

"I'll walk you out," said Putnam. "I've got something to ask you."

Ding…

"Wow. What's that? I almost forgot what the bell sounded like," said Cole, licking the last bit of chocolate frosting from his fork.

Ddddrrrriiiinnnngggg!

"Engine Sixty-one. Flooding. 7343 Fiesta. Cross Adobe. Fourteen thirty- two."

"Come on, squaddies," said Cole. "You better come along for manpower. This could be a water push."

The firemen hastily said their goodbyes to the Cookie Girl and parted company—each with a smile; each happy the reunion had gone so well.

Not using their lights and sirens, it took about ten minutes to drive non-code across town to the Adobe Tract, but that was fine with Terrence Putnam, the extra time allowed him to think of Evelynn and plan his next move.

Within several blocks of the reported address, Barker pointed to water running down the gutter. The runnel led them up the driveway and to the front door of a small, old house. The neighbor met them in the front yard and said the owners were out of town. Cole radioed the dispatcher and requested Truck 45 to respond with their water vacuums.

The firemen slipped into their rubber fire boots and looked for a way inside the house. They hopped the side fence and walked through the intricate maze of a beautifully manicured rose garden. All around the house, the curtains were drawn, making it difficult to see the severity of the flooding.

Putnam found an unlocked bathroom window, but only Valdez was able to squeeze through the tiny opening. Once inside, he waded to the front door and forced it open against the back pressure of the standing water, releasing a torrent of water and floating debris.

The front room was literally a sunken living room. All of the furniture was darkened from having wicked up the water. The floorlength drapes were water-stained halfway up and hung heavy, bowing their rods.

"Those damn couches will weigh a ton," scorned Barker, directly into Cole's ear.

Cole sloshed through the small house, plastic shoes and balls of lint bobbing in his wake. The sound of splashing water led him to the fractured toilet. The porcelain tank had a large crack on its backside. Cole reached for the shutoff valve. He wedged his hand along the back of the toilet tank, causing the entire corner to fall off releasing another torrent of water.

Pressurized water sprayed from the supply line into his face. He stretched over the bowl, fumbling for the shut-off valve. Contorted over the toilet, he twisted the handle and felt a similar twisting in his stomach. He gripped his abdomen and leaned against the wall as his stomach twisted into a knot. He grimaced and struggled to gain control of his bowels.

Shuffling down the hall and into the living room, Cole saw Valdez leaning against the wall. He looked pale. Cole needed fresh air, so he headed to the backyard. The sliding glass door was already open. He stepped into the cool air of the patio just in time to catch Barker leaving the rose garden and buckling his pants. It was out of character for the salty old engineer to blush, but his face was fire engine red. Normally, Cole would have said something jocose, but he was concentrating on his churning stomach as it audibly gurgled.

Truck 45 announced its arrival with the release of its air brakes. Cole shook off his discomfort, straightened his posture and went out to meet Captain Decker. Decker's firemen busied themselves around the ladder truck, pulling out the water vacuums, mops and squeegees. Captain Decker said nothing to Cole. He stood rigid and calloused as Cole described the flooded condition of the house.

Captain Decker ordered his firemen to shut off the electricity, remove the thresholds and the toilet.

"Why do we have to remove the toilet?" asked Cole, trying not to sound panicked.

"So, the standing water can escape directly down the drain hole," said Decker, indignantly.

Cole gasped to himself at the realization that the small house had no other bathroom. His stomach twisted and growled as his intestine chambered an explosive round in his colon. Cole quickly turned around, pretending somebody had called to him. "Okay, I'll be right there," he answered the phantom caller.

Captain Decker cocked his head, listening for the caller. Speaking over his shoulder, Cole told Captain Decker to continue and carry-on and so forth, as he scurried tight-kneed to the rose garden.

Picking up the water-soaked furniture by a crew with intestinal compromise was a risky endeavor. They moved slowly and deliberately, and took frequent breaks. Captain Decker eyed Cole's crew with sour contempt. It took two of Walker's firemen to carry a small upholstered chair only a few feet before stopping and sitting on it. Never before had Captain Decker seen such a lazy group of firefighters. He told Captain Walker so.

"They spend more time goofing off in the back yard than they do on the urgent matters at hand," snapped Captain Decker, not trying to hide his anger. "You were a slacker in the academy, Walker, and you're a slacker now," sneered Captain Decker, standing nose-to-nose with Cole, just like they were back in the fire academy. He took the same demeaning tone, with the same hot breath. "Why don't you go back to being a paper shuffler?" asked Decker. "You don't belong out here." Not waiting for a reply, Captain Decker pushed past Cole and marched into the house. Cole marched after Captain Decker, but turned left inside the house and headed back to the rose garden.

It took almost two hours before the water removal was complete. The water push dragged on for a variety of reasons. There was a lot of mopping, lifting heavy furniture and many trips to the rose garden. It became more and more difficult to find a decent area to admire the roses. Exhausted, sore and scared to move, the firefighters of Station 61 sat on the water-soaked furniture in the front yard. Without offering to help, they idly watched the firemen of Truck 45 reload their equipment and disappear down the street.

Although it was against department policy, Captain Walker ordered both engine and squad to use their red lights and sirens and respond code-three back to the fire station.

The entire crew had to go home for the rest of the shift. Cole was the last to be relieved. When he came downstairs after his brief shower, he met his relief captain who was dressed like an Indian Guide chief.

"My wife said it sounded urgent, so I came as soon as I could," he said, patting his cellular phone clipped to his loincloth.

"You don't know the half of it," said Cole. "Thanks for coming so quick."

"What do you suppose happened? Were you guys exposed to a flu-bug or something toxic?" asked the Indian chief.

"I'm not sure, but I suspect it was something we ate," said Cole over his shoulder as he trotted upstairs to the bathroom. While he sat, he planned his route home, careful to include all the possible locations for emergency stops. Cole readied himself and headed for his car.

"Cole, wait," called the Indian chief. "I found this slid under the office door," he said, waving a pink envelope.

Taking the envelope, Cole quickly tore it open. Inside was a greeting card with a couple of wild horses drinking from a serene mountain stream. Inside the card was a handwritten inscription:

Thanks for dinner. How'd you like dessert?
-Evelynn

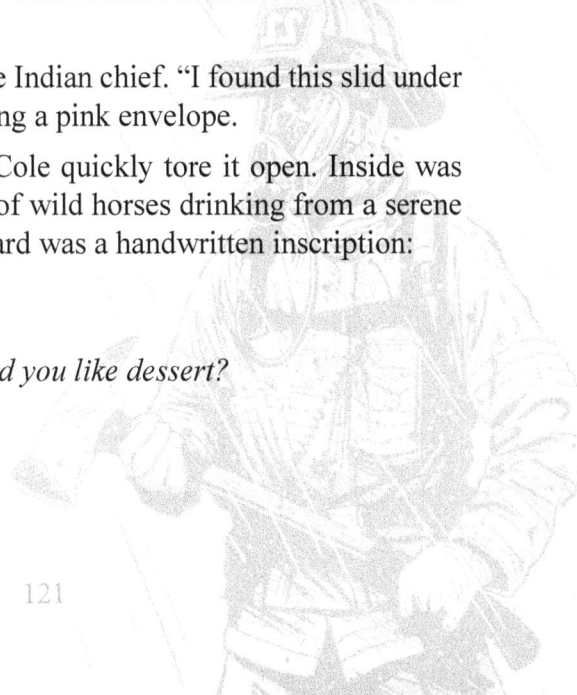

CHAPTER 10

HIDE AND SEEK

Careful not to make a sound, six-year-old Bradley sidestepped across the yard through the gauntlet of scattered toys. He could not afford to make a mistake. His opponent was good—very good. He pressed to the side of the house and peered around the corner. *Nothing.* Then he checked under the porch and found his two-year-old brother who was always the easiest to find.

"I see you," said Bradley, allowing his younger brother to scramble from his hiding place and race for the freedom of the clothes pole. The little boy's red sneakers slapped against the ground as he scampered past his older brother.

The two-year-old giggled and swung giddily around the pole, taunting, "Free. Free. I'm free."

"Wow, Jamie, you're fast. Now, I have to look for the other guys."

Freckle-faced Bradley continued his patrol for the other two boys hiding somewhere in the back yard. He looked behind the tool shed, over the hedge and under the boat. He glided past the pair of muddy shoes protruding from the lilac bushes. Four-year-old Tommy bolted toward the clothes pole, screeching the whole way, but his older brother did not give chase because he was searching for the real prize—his friend, Teddy.

Bradley grew frustrated because he couldn't understand where Teddy could be. It was a small yard. He had checked all the usual places. Some, he'd checked twice. When Bradley reached the far corner of the back yard, Teddy dropped from the neighbor's avocado tree and trotted easily to the clothes pole.

"Free! I'm free! And you're *It* again," Teddy shouted as he joined the younger boys' chorus, swinging around the pole.

"That's cheating," protested Bradley. "You were out of bounds. You can't hide in the neighbor's yard. *You're* It!"

The boys squabbled for a moment, but eventually conceded that their yard was too small and everybody knew all the good hiding places. So, they agreed that the boundaries would be expanded. Since Teddy lived only a few houses away, the boundaries were extended to include both their yards and the ones in between. They sealed their agreement with a spit handshake.

"But you're *It* because you were out of bounds," said Bradley.

The accusation was true, so Teddy squeezed his eyes shut and began counting into his arms folded against the metal clothes pole. "One one-thousand. Two one-thousand. Three one-thousand..."

The three brothers took off running down the block. Young Jamie trailed behind his larger brothers as they hunted for a hiding place. "Hurry up, Jamie," they called, waving for him to catch up. The little toddler's legs tried to keep pace, but they were not as strong as his willingness to participate. The two older brothers each grabbed an arm and hoisted little Jamie in tow as his little red sneakers paddled midair.

"How about there?" asked Tommy, pointing to a void behind the vacation trailer and the brick wall. "Or there," he said, pointing under the old Dodge pickup perched precariously on cinder blocks.

"Oh, I know a good spot," exclaimed Bradley, as he led his brothers to the perfect hiding place.

Spring debuted in Brookfield like it had for the past hundred years. Although there were no longer any rolling dairy pastures, new grass sprouted in the neat rows of residential lawns. Blooming flowers no longer blanketed the banks of the small creek that had meandered through town, instead blossomed from window boxes and neatly manicured flowerbeds. The air teemed with flickering butterflies and roving bees.

Everybody at the firehouse awoke to spring in his own way. Putnam noticed the birds each morning. Their delicate song reminded him of Evelynn. The clear blue morning sky reminded him of her eyes and the sweet fragrance of bursting hyacinths reminded him of her soft, freckled shoulders. Lately, everything reminded Putnam of Evelynn. He became annoyingly happy and took up whistling, which amounted to nothing more than blowing shrill air through pursed lips.

Annoyed by his hay fever and Putnam's pleasant disposition, Barker also noticed spring through his watery eyes as he picked smashed butterflies off the fire engine's front grill. Razmunson enjoyed spring because that meant hot summer weather was just around the corner, bringing women in halter tops and short shorts with it. The moderate spring climate made Valdez happy because winter rains and summer humidity flattened his hair. For Valdez, spring was good for another inch or two in height.

Spring also affected Cole Walker. His thoughts were filled with the continuum of life, in particular, cats and the people who love cats. He pondered the finer points of cat ownership, while trapped in a phone conversation with a lady near hysteria because her cat was in a tree.

The lady was frantic because Flossy, her 'baby,' had been up a tree for nearly three hours. She spewed details of the cat's dire situation, one sentence flowing into the next. When the cat-lady paused to catch her breath, Cole consoled her, but was quick to point out that placing a ladder in a tree was dangerous. He suggested leaving food under the tree and Flossy would come down when she got hungry. The lady insisted that the fire department retrieve Flossy immediately. Cole was forced to use the analogy firemen have relied on for eons.

"Ma'am, have you ever seen a cat skeleton in a tree?" His logic was undeniable. The cat-lady stammered for a moment as if punched in the head, then hung up the phone with a slam.

"Chalk one up for Captain Walker," said Cole to Captain Hastings' portrait.

He returned to his paperwork, but paused to visualize a cat skeleton in the bough of a tree. The phone rang.

"Captain Walker, I just received a call from my niece," said Mayor Jenkins. "She said you refused to get her cat out of the tree."

Cole covered the phone mouthpiece and spoke to the picture of Captain Hastings. "The cat-lady is the *mayor's* niece."

Mayor Jenkins continued, "Your public support is waning, Captain Walker." The mayor paused just long enough for his words to impact, but not long enough for Cole to retort. "A week ago, I received a complaint from a man who said you and your crew went to his house and shit on his rose bushes." The mayor paused to clear his throat, which serendipitously covered the sound of Cole slapping his hand to his forehead.

"I've been in public office for a number of years, Captain Walker. I know a prank call when I hear one. But what concerns me is that you have driven these citizens to such extremes. Do you like bad press? Does Chief Pierce enjoy bad press? I doubt it. Therefore, Captain Walker, I suggest you get off your damn butt and get the damn cat out of the damn tree!"

The phone line went dead. Cole shrugged his shoulders at Captain Hastings, "Can't win them all," he said, then assembled the engine crew and headed for the damn cat.

The tree with the stranded cat was easy to find. The ancient maple was on the corner lot directly behind the smug fat lady standing cross-armed in a pink housecoat. Both lady and cat glared at the fire engine as it drove up.

Bleve barked at the cat. The cat hissed at the firemen. The rotund cat-lady turned abruptly from the firemen and spoke to the cat in syrupy dialect and obnoxious kisses. Putnam and Barker marched to the base of the tree with a 24-foot extension ladder. The tree's flimsy canopy offered marginal ladder support. They raised the fly section through the web of branches, resting the tip near the angry, hissing cat. Putnam was assigned to retrieve the cat because he was the boot, and because Barker was too close to retirement.

Walker and Barker steadied the wobbly ladder as it fulcrumed against the spongy branches. Putnam ducked and wove like a prize fighter trying to out-maneuver the swatting, spitting cat. Cole and Barker had to use their full body weight to steady the ladder against Putnam's cantilevered mass. Their struggle was interrupted by the dispatcher's voice.

"Engine Sixty-one, L.A. What is your status?"

Cole reached for the radio, but the ladder fell away. He arched back and strained to stabilize the ladder. Another call came from the dispatcher. Cole held the ladder with both hands and had the cat-lady place the radio to his mouth and key the transmit button. He told the dispatcher that Engine 61 was available on their current assignment.

The dispatcher acknowledged and sent the following dispatch: "Engine Sixty-one, Squad Sixty-one. Public Assist. Lost Child. 2312 Poinsettia. Sixteen forty-seven."

At that same moment, Putnam made a grab for the cat. The catlady removed the radio mic from Cole's mouth about the same time the angry cat shook itself free from Putnam's grip. The frantic cat, falling in true form, landed claws first on the cat-lady's back. She leaped with a shriek. With one hand over her shoulder and the other back around her waist, she spun around and around.

As the robust woman spun faster and faster, her pirouettes became more elliptical. She dizzily stripped out of her housecoat, exposing a pink teddy, cadaverous white cellulose thighs and purpleveined legs. The housecoat bounded across the lawn with the cat-lady giving chase, thighs quivering with each stride. The firemen quickly replaced the ladder on the engine as the cat- lady grappled with her housecoat.

"I bet that cat would freak at a siren," said Barker, grinning impishly.

Cole gave the cat-lady and Flossy a courtesy wave before stomping on the air horn. The blast gave the cat the additional incentive to break free. Under the weight of the cat-lady's arm, the cat ran in place on her shoulder before breaking free and scrambling over the fence into the back yard. Cole waved good-bye to the cat-lady as she glared through her dangling curlers. He let the siren wail as they responded around the corner.

The siren's shrill sent the frenzied cat scrambling across the back yard, over the side fence, into the street and under the engine's rear tires. Flossy barely gave the engine a noticeable bump. Barker glanced at Cole who had his nose buried in the map book looking up their next address. Barker casually shifted into third and sped to their next call.

Engine 61 reached the incident the same time as the paramedics. Sheriff cars with red and blue lights spasmodically flashing were staged in front of the children's house. A helicopter circled overhead. Officer Parker was organizing the search and met Cole as he stepped down from the fire engine. "We have three boys missing for over five hours. All brothers," he said. "I'll need you to help us search."

As Officer Parker led Captain Walker to the command post, they passed the boys' mother. She trembled so much that she had to be steadied by a neighbor. Her eyes were red and swollen with tears. She clutched Cole's arm as he passed. "Please find my boys," she pleaded.

"We'll find your boys, ma'am. Don't you worry," said Cole. The woman collapsed into uncontrollable sobbing.

The command post was a map unfolded across the hood of Parker's patrol car, one edge anchored with a large coffee cup, the other with a box of donuts. Officer Parker circled the areas that he assigned to Cole. "Their mother says they never stray far," said Parker. "We want to make sure that they aren't in somebody's house watching the Flintstones before we call this a kidnapping."

Cole took the information and divided the search area amongst the crew. Each fireman was assigned a five-house section. The paramedics searched as a team, while Barker, Putnam and Cole split up to cover more area.

Barker began his search at the house with a pool. The house was unkempt and locked up. The lawn was overgrown and the hedges untrimmed. Barker walked across the oil-stained driveway, through the unlocked gate and into the back yard. The pool water was dark green and thick with algae and debris. Leaves and plastic shopping bags floated half-submerged, so he couldn't see deeper than an inch or two into the water. Grabbing the long-handled pool net from its hanger, he stabbed it into the murky water probing the bottom for what he hoped not to find. The disturbed water emitted the rancid stench of rotting pond scum. Chunks of black algae floated in the wake of the skimmer. Barker moved the pole slowly, deliberately, as he worked his way toward the deep end.

Meanwhile, Putnam walked through the back yards of the houses on the opposite side of the block. Each house had a woman home. No one saw anything unusual, and each offered to search their own property for the boys. There was little room to conceal three children in any of the neatly manicured yards and none had pools or outbuildings. Putnam walked through the rear of the last yard and looked over the weathered wooden fence and saw Barker probing the pool a couple of houses away. He called to Barker, "Need any help?"

Barker stopped his probing and looked around to see who was calling. He finally saw Putnam's bulbous head looming down the fence line. "Sure," Barker shouted, cupping his hands to his mouth; they stunk of swamp water. "Check these three houses," he said, pointing to the houses on his left. Putnam signaled with a thumbs up and looked for a way into the other yards.

Cole's search had proven fruitless. Most of the residents were home and very cooperative. The only house left was the most dilapidated. The front lawn was dead. The drawn window shades were old and stained. From the front porch, he could hear voices inside the house. He knocked on the door and waited. There was no answer. Looking through the small opening between the dirty window and the dirty sash, he saw the flickering light of the television. He knocked again—louder. There was movement inside before the door cracked opened.

The vile odor coming from the house smelled of dog feces, stale beer and marijuana. The inside of the house was dark. Cole couldn't see the person behind the door. He explained the situation with the missing boys but was interrupted mid-sentence with "Nope" as the door slammed closed. Cole ignored the resident's distemper and marched tight-fisted down the driveway toward the back yard.

The gate to the back yard was obscured with weathered boxes of glass bottles, bundles of yellowed newspapers and a variety of rusting car hulks. He stacked some boxes by the fence for a better vantage point into the back yard. The grass had long since lost its battle against the weeds. There were clear areas strewn with dog feces. In the far corner of the yard were a shovel and a fresh mound of dirt about four feet square. Cole vaulted the fence and headed for the earthen mound.

The stench of the pool's rancid water was suffocating. As Barker gave his last few sweeps across the deep end, he hit something solid. Something on the bottom of the pool made his pole bend. He forced the skimmer head underneath the unseen object and pried it up. The object slid begrudgingly along the bottom of the pool. Beads of sweat collected on Barker's brow and streams of putrid water ran down his forearms and dripped from his elbows.

Barker fumbled awkwardly with the pool skimmer, poking at the object, which felt solid, yet moved with considerable effort. He pressed the rim of the flimsy net against the object, slowly sliding the mass toward the shallow end of the pool. As he pushed the object along the rough pool bottom, he scanned the other yards and saw Putnam's head bobbing along the fence line several houses away. The aluminum rod arched as Barker carefully balanced the object on the skimmer net and forced it to the surface.

The green water made the skimmer handle slippery and difficult to grasp. With one hand, he quickly pulled out his wallet and eyeglasses from his pocket and tossed them onto the sun-bleached lounge chair beside him. Stinging sweat ran into his eyes, but he couldn't wipe his brow or he'd lose his grasp on the precariously balanced load. The slimy green water churned as blue and white striped fabric broke the water's surface. Barker sighed with relief at the sight of the water-soaked lounge cushion. He released it into the murky water; thankful he had not found the children.

Putnam examined the weathered fence and found a couple of loose boards that pivoted on their top nails, allowing him to squeeze between the slats and duck into the back yard of the next house. The yard was beautifully manicured with no clutter for children to hide in. He made a quick perusal and headed for the driveway. While passing the side door of the detached garage, he gave the doorknob a twist. It opened. The garage interior was dark and cool and smelt musty with a hint of chemicals. Putnam opened the door wider and called for the boys as he groped the wall for the light switch. "Tommy, Jamie, Bradley." There was no answer.

He swept the wall left of the doorway. He nudged the small refrigerator forward and feeling blindly, his fingers found the light switch. The fluorescent lights flickered and hummed before fully illuminating the room. Black and white photographs hung from overhead stringers and large posters decorated the walls. Apparently, the garage was the studio of an amateur photographer. He called again for the boys and stepped inside the garage until he could see to the far wall - nothing.

On the way out, Putnam squeezed his thick arm behind the small refrigerator and turned off the light switch, then he slid the refrigerator back against the wall. More out of habit than curiosity, he grabbed the refrigerator's handle and opened the door causing a dozen boxes of film to fall out and one small red shoe. His eyes widened in horror. The three little boys were wadded in the refrigerator's small interior. Putnam called for help while grabbing the first boy he could reach. The little boy's lifeless body molded itself to the contour of the refrigerator's stark square interior. Putnam ran outside, carrying the boy to the back lawn and yelled again for help. The boy lay scrunched and folded on the grass. Putnam blew into the boy's cold, blue mouth and felt for a pulse. There wasn't one.

Putnam ran to the back fence and screamed to Barker, "I found them! Get the squad! I found the boys!"

Putnam raced back into the garage and pulled a second little body from the refrigerator. He saw the inside walls were streaked with little bloody finger marks from the boys' desperate effort of claw their way out. He laid the crumpled little body on the grass next to his brother, but he too retained the cubical form of the refrigerator box. Putnam could not straighten the boy's arms or legs that were locked in morbid contortions. He returned to the refrigerator, frantically murmuring, "Oh, my God... Oh, my God..." He tugged at the last boy, the smallest of the children, who was wedged in the bottom of the fridge. His little face was scratched and imprinted with pants fabric from the pressure of his brothers' weight. His little red shoe fell to the floor, his socks hung halfway off. Putnam blew into each boy's mouth. Their lips were cold and leathery. Their eyes stared lifeless toward the sky. Putnam murmured between his ventilations, "Oh, my God... Oh, my God..."

Squad 61 screeched into the driveway. Razmunson ran to Putnam with his drug box and oxygen tank in hand. He watched Putnam frantically performing CPR on the rigid bodies and pressing futilely against their stubborn limbs to lie down.

"Do something!" demanded Putnam between his respirations into the unresponsive mouths of the little boys. His face was blood red and his murmuring had deteriorated into a chant.

Razmunson grabbed the wrist of the closest boy and instantly recognized the undeniably mottled skin and rigid joints of rigor mortis. The same with the second boy and the third. Razmunson turned his concern to Putnam who was desperately continuing CPR. "Terry, they're dead. There's nothing we can do," said Razmunson in a hushed voice.

Putnam continued his frantic respirations. "Try something, Raz," he pleaded. "Please."

"They're gone," said Razmunson, demonstrating his point by pressing on the boy's extended arm, unable to flex it. "They've got rigor, Terry. They're gone. There's nothing we can do." Razmunson stood and gently squeezed the big fireman's shoulder.

Putnam shook off Razmunson's hand and continued his resuscitation efforts several more times before he sank back on the grass and squeezed his legs against his chest. He looked at Razmunson for the first time. His eyes were red and teary. He shook uncontrollably.

Razmunson spread out his arms in a helpless pose and repeated with as much sympathy as he could muster. "There's nothing we can do. They've been down too long." He delicately pressed on the little arm extended defiantly to the sky, "They're gone."

Officer Parker and Captain Walker and a horde of searchers ran up the driveway and stopped silent when they saw the stiff, contorted bodies of the boys. An eerie silence enveloped the crowd. Cole spoke to Officer Parker without breaking his stare at the little bodies. "Get some blankets, Stan. Don't let their mother see them like this." Officer Parker just nodded and went to his patrol car for blankets.

Cole's eyes moved to Putnam who was rocking back and forth, his knees squeezed tight against his chest. Lifting gently on the big man's arm, Cole said, "Come on, Terry. Let's go home."

The ride back to the station was long and uncomfortably silent. For the remainder of the day, the firehouse remained unusually quiet. Barker sat in front of the television and mindlessly flipped through the channels. The paramedics were upstairs. Valdez was on the weight bench. Razmunson pushed hard on the stationary bike. Alone in the front office, Cole could not concentrate on his report. The more he thought about the incident, the more vivid the vision of the three boys flooded his thoughts. He set his pen down and massaged the bridge of his nose. He rested his head in the palms of his hands and lowered his head onto his desk.

A soft knock came at the office door. Cole raised his head and checked his face for creases in his reflection on Captain Hastings' portrait. He gave his face a vigorous rub before answering the knock. Putnam stepped inside the office and pulled the door closed behind him. His shoulders slumped and his head hung heavy as he stared at his feet. When Putnam eventually raised his head, Cole could see that he had been crying. Putnam looked at the ceiling, then at the floor, then out the window, never making eye contact with Cole.

"What's on your mind, Terry?"

The big man stammered and swayed, obviously struggling to produce words. He cleared his throat several times. "Captain Walker, I've been doing some thinking and I don't think I'm cut out to be a fireman," he said. "I don't think I'm right for this job," he said as his voice rose in pitch. "I tried to help those kids…but I couldn't." Putnam stopped speaking and struggled to control his quaking shoulders.

Cole stood and slid his chair to Putnam. The oak chair creaked as the big man sank into it. Swiping his big hand across his face, Putnam wiped his eyes, then his nose, then his eyes again. Cole looked for a tissue but could only find a sheet of typing paper. He put it back. Putnam spoke into his fidgeting hands folded in his lap.

"I don't think I'm meant to be a fireman," he said, not bothering to wipe his tears anymore. Looking up for the first time, his chin wrinkled and trembled as he continued, "I think I should quit." Hanging his head once again, Putnam pressed his large hands to his large face and cried.

"Terry, Terry," said Cole, sliding off the desk and kneeling next to Putnam. "You didn't do anything wrong. You couldn't have helped those boys. None of us could. It was too late." Cole put his arm around the big man's trembling shoulders.

"I can't do anything right. I think I'm responsible for Captain Hastings' death," he said. "I can't shake the vision of those little boys. For everybody's safety, I think I better quit. I'm a jinx."

Cole's throat tightened and burned. He tried to clear his throat, but it was too dry. "Terry, you're the most sensitive fireman I know. That's a difficult thing to be in this business. Don't be like the other firemen who get thicker skins with each incident. You're not like them." He lowered his head to make eye contact with Putnam but could only see his face buried in the palms of his thick hands.

Putnam kept sniffing and wiping his eyes, but his efforts could not keep up and his tears fell to the floor.

"Terry, you have experienced more in your first year than most firemen see in a career, but don't be confused with what happened and what you could have changed. You couldn't have saved those boys. Time was against us. It's not your fault."

"But I can't get them out of my mind," sobbed Putnam. His round face, red and wet with tears, lifted up to Cole. "I can't forget their faces… their little bodies…"

Cole put his arm again across Putnam's wide shoulders. "Terry, you're not supposed to forget those boys. Remember them. Because someday, you'll be a father. And you will be a good father because of those boys. You will double check and triple check on your children. You will worry when they're out of sight and you will watch them and constantly check on them because of those boys. And when you return home after work, you'll hug your children and kiss them and know just how precious they are because of those boys." Cole's voice cracked. He paused for a moment and gently squeezed Terry's shoulders. "Terry, don't quit. There are more boys out there who need you. *We* need you."

Cole's throat was too tight to continue. His words hung in the air. Together, the two firemen hung their heads in the silence of the office and wept.

CHAPTER 11

HAY DAY

Saturday morning showed promise of becoming another beautiful day. The morning was already bright and warm inside the captain's office. Small dust particles swam in the sunlight streaking through the office window. Cole grabbed the station calendar and rechecked the date against the flyer in his hand. In bold black letters, the heading read, *Hay Day Celebration and Parade*. The artist's conception on the poorly reproduced flyer depicted a marching band, a clown and a fire engine – none of which had anything to do with the original holiday. The original Hay Day commemorated the day when the Brookfield dairy farmers filled their barns with enough hay to last the winter. The farmers took the following day – or week – off from their chores to nurse their tired backs and throbbing hangovers.

As civilization invaded Brookfield, the dairy farmers disappeared along with their hay. The original date of Hay Day was lost. It was changed several times to dovetail with school vacation and the more prominent national holidays. Nothing remained of the original Hay Day other than the namesake and the next day hangovers.

The true origin of Hay Day was far from Cole's mind; he was just happy to be a parade participant and was eager to tell the crew. As he stepped into the kitchen, he asked Putnam to summon the crew for lineup. Putnam playfully saluted, marched into the center of the apparatus room and relayed the announcement at the top of his lungs.

Standing in front of the coffee maker, his ears still ringing from Putnam's line-up call, Cole pulled his coffee mug from the cupboard and rolled it in his hands, checking the rim for boot polish and the handle for honey. Soon the door was pushed open by Barker's butt as he backed into the kitchen, rubbing his greasy hands with a shop rag. Putnam followed close behind, stopping short at the refrigerator. Valdez and Razmunson bumped shoulders as they simultaneously tried to squeeze through doorway.

Routinely, morning line-up began with catch-up conversation covering what happened during their two days off. Barker was still remodeling his bathroom. Valdez had dated twins. Razmunson spent his time rollerblading on the boardwalk at Venice beach. Putnam said nothing, devouring some leftover potato salad and enjoying everyone else's stories. Cole was glad everybody was in such a good mood; the parade would be a nice treat. He beamed with boyish enthusiasm as he announced the parade.

"Oh, is that today?" scoffed Valdez. "How come the parade always falls on the C-shift?"

"Well, at least we aren't taking the Lady's Ugly Auxiliary… are we?" asked Razmunson. "Last year, I nearly got a hernia helping that big broad onto the hose bed."

"Mrs. Broadenhoff?" asked Cole.

"That's the one," said Razmunson, shaking his shoulders with a shiver. "When I was helping her over the hose bar, she tried to step on my face. Then the wind caught her dress and threw it over my head. I was trapped under her skirt with her brontosaurus thighs and old lady underwear. I'll have that mental scar for life."

Cole felt his cheeks beat with flush as the firemen railed the parade. He tried to sell the event as fun but was quickly rebuked by Barker.

"What's so fun spending three hours in the stagnant heat, smelling diesel exhaust and boiling radiator fluid?"

The way Barker put it, the parade didn't sound like much fun, but Cole hoped this time would be different.

She shut off the water and pulled back the shower curtain. She arranged the delicate undergarments draped over the shower rod, then studied her obscure image in the steamed mirror. "Not bad for a divorced mother," she said to her frosted reflection as she slapped her firm thighs. She slid into her bathrobe and lit a small incense candle on the bathroom counter.

She held a match to the stubby candlewick. Most of the candle was gone, melted into a thin slab of wax spread across a small paper plate. Pressed against the match, the wick flickered and grew to a yellow flame. The sweet redolence of kiwi and strawberry began to overpower the bitter sulfur of the smoldering match. She made a mental note to put *candle* on her shopping list; this one was almost gone.

Next, she opened the closet and rifled through her clothes. She slid hanger after hanger across the rod until selecting the appropriate outfit for the Hay Day parade, khaki shorts and a white cotton blouse. It was supposed to be a warm day, so she would dress light. She returned to the bathroom and crouched in front of the small section of mirror that the candle had cleared of condensation. Leaning closer to the mirror, she smiled and scratched at the speck on her front tooth.

"Ouch!" she exclaimed, jumping back and rubbing her elbow. She had leaned too close to the candle flame. "Doggone you!" She shoved the candle to the end of the counter, then returned her attention to the mirror where she primped her hair and applied makeup. Humming merrily, she rolled the tails of her blouse and tied them in a knot just above her navel, then spun from side to side, watching herself in the mirror, sliding her hands down her tight shorts which adhered to her well-contoured figure.

She tiptoed across the living room to the baby blanket mounded on the couch. She gently tucked in the edges and gave the blanket a delicate kiss. Backing out quietly, she softly closed the front door, planning to return before she was missed. As the door eased shut behind her, its light draft ruffled the window curtains and wavered the flame of the bathroom candle.

At the staging end of the parade route, Engine 61 and Squad 61 sat idling. "These things never start on time," scowled Barker, drumming his fingers on the steering wheel. He watched the parade directors scramble around, waving their clipboards, trying desperately to organize the inattentive participants.

The members of the Brookfield High School Band tooted and tuned their instruments in piercing disharmony. The drummers' tapping and the woodwinds' squealing irritated the skittish mounts of the Palomino Pals positioned behind the band and in front of Engine 61.

Cole studied the jittery, yet beautiful Palominos. He admired their shiny blond coats and intricately braided manes. "You could eat one of those?"

"Never again," said Rus, and they both shared a moment of laughter.

Cole enjoyed Rus' laughter. He did it so rarely. Maybe, deep down he liked parades?

The woman in the cotton blouse made a U-turn on an obscure side street and parked her car near the corner. She hoped for a fast exit after the parade because she needed to get home before feeding time. Looking up the block toward Main Street, she could see people gathering, but there were still some choice spots by the reviewing stand. *Perfect.*

The parade directors worked their way through the staging area, gradually forming structure to the clot of parade participants. One director, passing by Engine 61, climbed on the running board to speak to the captain. Since Engine 61 was positioned directly behind the Palomino Pals, the director asked Cole not to use the siren or air horn because it would agitate the horses.

"Putting a fire engine behind a pack of jittery horses is dumb planning," vented Barker.

Cole agreed. "What kind of an impression are we going to make if we can't run our siren?"

From what Putnam could see from standing on the fire engine's motor cowling, the parade was starting. First, the Brookfield sanitation truck laced with red, white and blue ribbon rolled forward. Then, the hay wagon loaded with Brookfield High School Cheerleaders jerked into motion, knocking them down into contorted positions and causing the crowd to cheer. Next, the Shriners, chasing one another in midget cars, spiraled down the route. The Brookfield Mortuary Motorcycle Escort Corp, stacked in a human pyramid, motored behind the Shriners. Bleve barked with excitement, perking the ears of the nervous Palominos. Finally, the Brookfield High School Band stepped into formation. The Drum Major blew his shrill whistle and pumped his baton starting the band in motion.

The thin paraffin dam of the cratered candle gave way to the heated pool of wax. A stream of molten paraffin flowed as a miniature river from the paper plate onto the countertop, quickly exposing more candlewick. The candle's flame gorged itself on the exposed, cantered wick and, growing fat, impinged directly on the paper plate.

The drum major blew two quick tweets from his whistle, causing the Brookfield High School marching band to explode into an enthusiastic rendition of *Stars and Stripes Forever* as they began marching down Main Street. Following close behind, the Palomino Pals lunged forward on their skittish mounts. Barker pulled at the knot in his tie and mopped his brow before putting on his dress cap. He released the air brake and relaxed the clutch pedal. The fire engine slowly rolled forward, creeping behind the unruly ranks of the nervous horses.

Early in the parade, out Cole's side window, a beautiful brunette in a Hawaiian blouse waved vigorously and blew kisses at the firemen. From the paramedic squad directly behind the engine, Valdez took notice of the woman's exploits. He inched the truck forward until the woman came even with the cab. He batted his Neanderthal brow, pursed his oversized lips and blew a kiss to the woman, which apparently struck her hard, knocking the smile from her face.

On Barker's side, a pudgy young boy waved frantically at the fire engine. The boy motioned up and down with his fist, the universal sign for an air horn blast. Barker helplessly shrugged his shoulders and said, "Sorry, no horn today," but the young boy kept pumping his fist.

Barker turned to Cole and yelled over the clamor of the parade, "The best thing about parades is nobody can hear a thing you say. Watch this." He nodded to the enthusiastic young boy still pumping his fist. Barker waved back and spoke in a volume that made Cole nervous.

"Hey, you little rat. Lay off the cotton candy. It's going to your hips." The boy's smile widened, and he pumped his arm with even more vigor. Deviously grinning, Barker turned back to Cole. "See, they can't hear a word."

Throughout the parade, Barker's slandering continued. For once, Russell Barker was enjoying himself, waving from his window and hurling his insulting commentary to the spectators. To the lady with the huge thighs ballooning out her stretch pants, he said, "It looks like you're smuggling 50 pounds of chewed bubblegum." At a man's toupee, he said, "Nice rug," and flashed the man a thumbs up. "It looks like roadkill." Barker's entertaining slander was infectious.

Growing comfortable with the realization that nobody could hear Barker's insults over the parade noise, Cole allowed himself to snigger with each malicious comment, further inciting Barker to interact with his two-mile audience. Despite the sadistic motivation for his pleasure, Barker was sporting a smile. He looked happy and much younger. Cole didn't see much harm in the activity, so he allowed Barker his diversion, and eventually he also joined in the lambasting.

"Nice pants…Going Golfing?"

"Everyone with B.O. wave."

"When does Bozo want that shirt back?"

"I can smell your breath from here."

"I've got one word for you—liposuction."

"That comb-over isn't fooling anybody."

The parade took on a new purpose. Engineer Barker and Captain Walker were no longer just stanch members of the fire department, but highly animated, friendly neighborhood firemen – or so it appeared. From the perspective of the parade spectators, they saw a happy, enthusiastic fire company. From the cab of the fire engine, the firemen saw a target rich environment.

The edge of the paper plate yellowed and curled as it neared its ignition temperature, then burst into a ring of fire. The escaping wax no longer hindered the flame, but allowed it to grow larger and taller, burning a circle into the countertop. Streamers of black soot lofted from the flames as they stretched to touch the decorative tassels of the towels hanging directly above.

The ornate hand towels, with their tassel ends and delicate lace borders, required little heat to reach their ignition temperature. The tips of the tassels momentarily glowed with small embers before birthing flames. The flames, only briefly small, raced up the towels, consuming them as well as the plastic towel rod. Quickly, the melting rod sagged and gave way to the weight of the burning towels. One towel dropped back onto the counter, igniting a box of tissues. The second towel slid off the angled towel bar and dropped into the wicker waste basket.

The fiberglass countertop blistered and popped from the heat of the rampant flames. Fire engulfed everything on the counter and spread uninhibited from the hand towel to the plastic toothbrush holder, to the box of cotton balls before the weakened countertop collapsed into the storage cabinet below.

Rolls of toilet paper, plastic hair curlers, boxes of tissue, and feminine products stored inside the cabinet fed the flames. Hot smoke and embers rose and mushroomed against the bathroom ceiling as the fire gutted the cabinet. Flames from the waste basket climbed up the bathrobe hanging from the coat hook and extended the fire to the delicate linens draped across the shower rod. Radiant heat of the unchecked flames ignited the window curtains. Intense thermal energy delaminated the cabinet's particleboard frame, releasing flammable gases that ignited with explosive force blowing fire throughout the bathroom.

Flaming pieces of window curtain floated through the air, landing in the dried flower arrangement on the opposing wall. The shower curtains fell in a flaming heap, igniting the wicker laundry basket. Flames from the inferno lapped out of the bathroom doorway and into the hall.

"Nice hair color. Is the circus in town?"

"Lady, you're huge. Eat more salads."

"Her mustache is bigger than yours."

"Who did your hair? Hurricane Andrew?"

"You otta sell shade."

"Nice beer belly. You must drink straight from the tap."

"Hey, Rus," said Cole, tugging on Barker's shirtsleeve. "Look at the redhead!"

Cole waved at the woman but tried not to look too conspicuous as he spoke out of the side of his mouth. The object of his attention was a slender, well-proportioned woman in her late thirties. Her huge, untethered breasts pressed firm against her white blouse like trade winds billowing a spinnaker. Each bosom bounced with excitement as she waved to the firemen.

Cole waved. Barker waved. He also was enrapt by the woman's delightful proportions. In his newly found youthful-jubilance, Barker tried to see more of the woman as she passed slowly across Cole's window. He had his hand on Cole's knee as he leaned hard right.

"If she keeps churning like that, she'll turn into butter," said Barker, pressing harder into Cole.

"In that case, I'm giving up margarine," said Cole.

Barker stretched across Cole to catch the last fleeting glimpse of the woman before she passed from view. With each enthusiastic bounce, the lady's bosoms gyrated in opposing directions. Barker brought the phenomenon to Cole's attention as he spoke with childlike excitement.

"Look! They're counter-rotating!"

"I can feel the breeze from here."

"She's ready for take-off."

"I hope she lands over here."

Barker's brain raced for more comments. He was on a roll. They were on a roll. The fire engine was on a roll and continued to roll toward the ranks of the stalled Palomino Pals. Still craning for his last look, Barker shifted his weight to his elbow then momentarily leaned his elbow on the air horn. The unnerved Palominos leaped into the last three rows of the Brookfield High School Band brass section, which in turn lunged left into the curbside spectators and right into the reviewing stand.

Mayor Jenkins was one of the judges. It made quite an impression.

Flames bit into the bathroom jamb and curled the thin veneer of the hollow core door. Over-pressurized cans of hairspray BLEVE'd in the bathroom, blowing flames and embers in every direction. Thick black smoke and scorching temperatures blasted into the hallway. Fire lapped across the hallway and into the opposing bedroom. Red, orange and yellow flames crawled across the ceiling, pushing the dense smoke down the length of the hall into the other rooms of the house.

Flammable gasses, carbon monoxide and other toxic products of combustion collected at the ceiling. The impervious black smoke banked down to meet the fire. The bedroom flashed with a roar as it exploded into a blazing inferno. The house echoed with crackles and crashes as ceramic lamps shattered and light fixtures fell to the flames. As if in agony, the house creaked and moaned as it succumbed to the power of the fire.

Cole walked as if he was in a trance in the middle of the chaotic parade scene. He sidestepped between the frenzied horses and the tuba player floundering in his tweaked instrument. Cole heard the alarm tone on his portable radio. He held the radio to his ear and listened to the call.

"Engine Sixty-one, Engine Forty-five, Engine Thirty, Truck Forty-five, Squad Sixty-one. Residential structure fire. One Seven Four Two Lime Place. Cross Peach. Fourteen Forty-eight. Multiple calls."

Cole scanned the tumultuous crowd for his crew. Putnam was struggling to free a corpulent boy from his bass drum. The paramedics were surveying the crowd for injuries. Luckily nobody was hurt. Cole whistled and waved to the firemen, sending them running back to their rigs and hopping into their turnout clothing. Spectators cheered from the reviewing stand, marveling how the performance looked like an actual catastrophe. From inside the engine cab, all Cole could see was chaos; bucking horses, people applauding and a tuba player reeling to free himself from the grip of his deformed instrument.

Cole stood on the siren button, while Barker pounded the air horn. The Palominos bucked and galloped in every direction. The spectators made no attempt to disperse. Instead, they applauded the lifelike exhibition. The fire engine's siren could not clear a path through the crowd, but the wild Palominos did.

"There!" shouted Cole, pointing to a gap in the crowd. The Palominos jumped and spun, scattering the spectators, enlarging the hole and revealing the entrance to a side street. Cole jumped from the fire engine and ordered the sea of people to part. The people slowly collected their chairs, blankets and ice chests, then shuffled aside allowing the engine and squad to squeeze off the parade route.

Barker pulled hard on the steering wheel but was unable to make the sharp turn onto the side street. Jockeying back and forth, five or six times, Barker got the fire engine pointed in the right direction. With siren's wailing, the engine raced down the narrow street.

Reaching the main boulevard, Engine 61 skid to a stop. The cross traffic from the exiting parade watchers sat gridlocked, blocking their egress. The firemen sat helplessly at the intersection. In the clear sky beyond the stalled traffic, a column of smoke billowed above the distant houses.

Cole stood on the air horn and siren buttons with both feet. Nothing moved. The vehicles directly in front of the fire engine were wedged bumper- to-bumper. The cars in the intersection a half block away heard the sirens and held their positions, patiently waiting at the green light and anticipating an emergency vehicle to pass by.

"Move out of the way!" screamed Barker, waving his arm back and forth at the cars in front of the fire engine. He slammed his hand on the horn button. "Move!"

Some cars inched forward, and others jockeyed backward, tightening their ranks and creating a narrow gap in the gridlock. The engine squeezed between the cars, only to stop against the street's center divider. Engine 61 sat sideways in the boulevard wedged between cars on each side and the median curb in front. Cole knew they were losing critical time. The smoke header was getting larger and other incoming fire units were getting closer.

"Do not let Captain Decker beat us in. No matter what!" Cole shouted at Barker.

"No matter what?' Is that an order?" asked Barker.

"That's an order!" ordered Cole.

Following the Captain's order, Engineer Barker shifted the transmission into low gear—*granny gear*. The motor revved and the clutch spun as the front tires bumped over the six-inch median, plowed through the decorative hedge, then bounced into the opposing traffic lanes. He eased the fire engine over the median until the rear wheels dropped onto the pavement with a decisive bounce. Squad 61 followed the engine's swath over the crushed shrubbery.

Their sirens wailed and the air horn bellowed as they accelerated toward the fire. In the distance, Cole studied the column of dark brown smoke. Brown smoke meant the fire had not yet burned through the roof. Cole knew if the structure was not fully involved, they still had a chance to save it. Knocking on the back window, Cole pointed to the smoke column, then held his hand to his face. Putnam signaled with a thumbs up and readied his breathing apparatus.

Cole squeezed the radio microphone, "L.A. Engine Sixty-one. Two minutes out. Smoke showing." The dispatcher repeated the message to the other incoming units. Squad 61 swerved momentarily as Paramedics Valdez and Razmunson hooted and slapped palms.

The traffic lightened farther from the parade route. The two fire units screamed up the street after changing back into the correct lanes of traffic. As they roared past the row of cars pulled to the curb, Engineer Barker was fixated behind the wheel like a gargoyle.

Intense on upcoming traffic, he shifted smoothly, unconsciously. He remained focused on his driving, oblivious to everything else including his long strands of comb-over hair flapping at full length out his side window.

One minute out. At the opposing end of the boulevard were the flashing red lights of Engine 45 and Truck 45.

"Will we beat them in?" asked Cole, staring at the approaching beacons.

"It'll be too close to say," shouted Barker, not taking his eyes off the road.

"Where do we turn?" asked Cole.

"Two more blocks," said Barker, he pressed harder on the accelerator.

Engine 61 quickly approached the second intersection. With little braking, Barker spun the steering wheel hard right onto Peach Street. The tires squealed over the siren's cry. Cole clutched his door handle to keep from sliding off his seat. They were ahead of Decker by half a block. The next left turn revealed the burning house.

The street was filled with people. Clusters of neighbors and children on bikes scurried from the street as the fire engine squealed around the corner. The burning house came into full view. The smoke helix was as wide as the house and was fed from every window and vent opening. The fire crackled and spit. From inside the house came a plethora of sounds; crashes from bookshelves and cabinets collapsing, windows breaking, and delicate tinkling as glass and ceramics shattered from the heat. The bathroom window exploded. Flames roared as they lashed out and clawed at the wooden eaves, then leaped into the column of churning smoke.

Captain Walker yelled into his radio, speaking fast and excited. "L.A., Engine Sixty-one at One Seven Four Two Lime Street. Single family residence well involved!"

While the dispatcher repeated his report over the fire-command frequency, Cole gave assignments to the incoming units over the tactical channel. "Engine Forty-five, take the hydrant. Bring us a supply line. Truck Forty-five, ventilate." He paused for a moment, allowing the units to acknowledge their assignments. He jumped down from the cab, grabbed Putnam by the sleeve and ordered an attack line through the front door. Cole then ran to the squad and yelled to the paramedics over the roar of the fire, "You two are the Rescue Group. Look for victims. Take a reel line with you for a lifeline." They acknowledged with nods and hastened to don their air bottles, then pulled the smaller reel line toward the house.

Cole returned to the fire engine and picked up the tactical radio. He made additional assignments to incoming units for exposure protection and utility disconnect. Cole also requested sheriffs for crowd control. He knew the media would be here soon.

His breathing apparatus strapped to his back, Putnam flaked out the hose line on the front lawn. Raising both arms above his head, he signaled Barker for water. The hose swelled and snaked on the grass as it hardened from the pump's pressure. Putnam shot a stream across the lawn then twisted the nozzle changing the water pattern from a straight stream to a wider, more protective fog pattern.

Putnam shook the front doorknob. It was locked. He pulled off a glove and felt the door with the back of his hand. It was relatively cool. With one seemingly effortless kick, the door gave way. Smoke blew from the opening, obscuring Putnam from view. He crouched down and studied the house interior. The smoke level was three feet off the floor and pressing lower. He tightened his helmet strap and paused habitually, waiting for a reassuring nudge from Captain Hastings, which didn't happen. He entered the house alone.

The living room, den and dining area were joined, creating a large open space. Black smoke obscured the high ceiling. Putnam hunched below the smoke and advanced farther into the house. A river of red-orange flame flowed from the top of the hallway like an upside-down waterfall cascading up into the den's vaulted ceiling. He shot water at the flames. The flames retreated from the cooling spray but returned with fury. He had to go deeper into the house and find the source of the fire.

Truck 45 rumbled up the street and parked directly in front of the house. Cole ordered Captain Decker to vertically ventilate the house. Captain Decker stood rigid beside the ladder truck and barked orders at his firemen to ventilate the roof. He watched his crew lean the 24-foot extension ladder against the roof eave. The truck engineer test-ran the chain saw, then handed it to the truck fireman already climbing the ladder.

"Hurry up, ladies," yelled Captain Decker, his voice booming over the noise of the fire. "We haven't got all day." The firemen scampered up the ladder, followed by Captain Decker.

Putnam pulled extra hose and racked it across the den floor, careful not to loop it over itself. He didn't want to crawl in circles retracing his hose line should he have to make a hasty retreat. He tugged on the hose and advanced down the hall as flames churned overhead. He opened the nozzle into a circular spray pattern and trudged down the hallway pushing back the flames. Mid-hall, he found fire blowing from two opposing doorways. He turned right into the bedroom awash with fire and shot water into the room. He attacked the flames hiding in the mattress, dresser and closet. But the fire in the bathroom behind him was sucked into the vacuum caused by the water spray. Heat and flame came from behind and blew all around him. He fell to the floor, searing flames biting his back; his shadow pulsated and twirled on the hellish lit carpet. He rolled to his side and shot water at the wall of fire, holding the flames in check. He advanced into the bathroom, pressing forward, dousing the fire as it lashed at him. He reached the inside corner, shut off the nozzle and studied the blackened room. Only smoldering objects lie before him.

Cole felt a twinge of anxiety, knowing that Putnam was in the house alone. He was glad Valdez and Razmunson were going inside with him. Cole took a deep breath and reviewed his fire attack strategy. Everything seemed in order. He was pleased with his equanimity. He fine-tuned his operation by giving additional assignments for garage protection to the west and salvage operations inside. The dark smoke issuing from the bathroom window turned white with steam as Putnam attacked the fire. Cole released his pent breath, relieved. Everything was falling into place. This was a textbook fire. Everything was going as planned.

"My baby! My baby!" shouted a frantic woman as she ran back and forth in front of the burning house. She shook Cole's arm as she pointed to the flames. "My baby is inside the house!"

Cole whistled to Valdez and Razmunson poised at the front door and held up his fist signaling them to hold.

"Where in the house is your baby?" he demanded.

"I…I think he's in the front room," she said, swiping at her tears. "He was sleeping on the couch."

Cole ran to the paramedics. "There's a baby inside. The lady thinks he's in the front room on the couch. Find him." The paramedics nodded and disappeared into the smoke.

Engine 45, laying a four-inch hose line from the hydrant, pulled up behind Engine 61. The metal hose couplings clacked hollow on the street as they paid off the engine's hose bed. Cole gave Engine 45's captain a briefing on the current fire assignments and the status of the fire. He said he had men inside searching for a baby. Cole passed his command to the other captain, donned his BA and followed the hose lines into the house.

Kneeling inside the front door, he took a moment to study the interior of the structure. The upper portion of the room was filled with an inversion layer of thick, dark smoke. The implacable, carcinogenic smoke loomed halfway down the walls, allowing him moderate visibility if he walked stooped. He could see the attack line trailing down the hall. The smaller reel line went left, into the living room. Cole saw the shadowy figures of the medics through the smoke's haze as they searched the living room for the child. He watched them pat chairs, probe under the couch and behind the stereo cabinet. He ventured deeper into the house, planning to search the rear bedroom for the baby.

Cole followed the attack line as it turned down the windowless cave like hallway. The smoke level in the hall was to the floor, forcing him to crawl on his hands and knees as he followed the hose line deeper into the darkened ingress. The outside of his face piece clouded with condensation as steam drifted from the fire and cooled against his mask. He wiped at the condensation with his gloved hand, streaking his mask with water and soot.

To his left, he noticed a small puppy, probably a Golden Retriever, sprawled quivering against the wall, victim of carbon monoxide and other poisonous fire gases. He did not pause, but only thought of the baby. To his right, coming into view through the haze, was Putnam spraying water into the bedroom. He hoped the baby was not in that room. If it was, he prayed Putnam would not be the one to find it.

Acknowledging Putnam with a pat on the back, Cole continued to the other bedrooms at the end of the hall. He heard the Truck firemen overhead. Their pike poles made dull thumps as they inched along the roof probing for weakened rafters. The chain saw powered up as the crew prepared to cut a ventilation hole in the roof. Cole continued his search.

The smoke level in the master bedroom forced him to crawl lower. He worked his way around the king-size bed, feeling above and beneath it. From his crouched position, he could distinguish only the faint outlines of the furniture. Overhead, the noise of the chainsaw was replaced with the muffled thudding of pike poles pounding through the ceiling drywall signifying the completion of the ventilation hole and the release of smoke and gases to the outside. Cole noticed a crouched fireman to his right.

"Check the back bedroom," he shouted, muffled by his face mask. The fireman did not move. "Check the back bedroom," he repeated louder, pointing back to the hall.

The fireman mimicked his indication to the hall. "Right," Cole acknowledged. The fireman still did not move. Apparently, the fireman didn't understand Cole's order. The two firemen crawled closer toward each other until Cole bumped his face mask into the wardrobe mirror. He checked quickly for witnesses and resumed his search.

The next noise to pierce the smoke was the smoke blower. The large fan was positioned at the front door, pressurizing the house with fresh air. Almost instantly, visibility improved as the smoke was blown out of the hole in the roof. Cole was able to continue his search on foot.

Razmunson and Valdez met Cole in the master bedroom. Their search for the baby was unsuccessful. Cole sent them into the back bedroom and hoped the baby would not be found in the fire-ravaged bedroom. Shortly after they left, the two medics returned and reported that they had found nothing. Cole's heart swelled into his throat, as he feared the worst. *The child must be in the fire room.* He headed outside, praying he would find mother and baby safe.

On his way out, he checked on Putnam who was in the closet blasting a straight stream into a lump of smoldering clothes. Smoke still issued from the doorframe and window casement. The entire bedroom was black and gutted, the walls mottled in hues of black and gray. The floor was littered with mounds of charred, steaming debris. Cole didn't tell Putnam about the missing baby, he wanted to check with the mother one more time.

Cole knew he would need the medics close by when he told the lady that they could not find her baby. Still clad in his BA, with face mask muffling every syllable, communication was difficult at best. Cole augmented his words with hand signals, waving for the medics to follow him outside. Valdez and Razmunson trailed him down the hazy hallway toward the beacon light of the front door.

Stepping onto the front stoop, loosening the straps from his helmet and smudged mask, Cole scanned the crowd of onlookers for the lady in the white blouse. There, with Officer Parker, she stood empty handed. Hellish thoughts flooded his mind. *What will I say? How will I tell her that we didn't find her child? What will she do?* He walked directly to the woman, fearing eye contact. She turned from Officer Parker and stared directly at Cole. Her face was pale, her eyes swollen and flowing with tears.

"My baby!" she screamed, startling Cole. Stretching out her arms, she ran past Cole to Razmunson who was carrying the limp Golden Retriever. Its coat was matted with water and blackened with soot. The woman stroked the unresponsive puppy. She pressed her cheek against its matted fur and caressed the dog's lifeless body still cradled in Razmunson's arms. Looking up tearfully, she pleaded, "Can you do anything for my baby?"

Cole consoled the woman and nodded to Razmunson. The paramedics ran the dog to the paramedic truck and began CPR with mini compressions and an infant oxygen mask. Ignoring the flashes from the news cameras and reporters' questions, the paramedics worked on the flaccid animal. As Cole watched the medics work on the dog, he was spun around by the shoulder.

It was Red Hall. Cole was so happy to see Red that he spontaneously, uncontrollably gave him a hug. Red returned the favor.

"Hey, Red," said Cole, shaking Red's hand so hard it made his helmet rattle. "Want a good humanitarian PR piece?" He motioned Red toward the paramedics and the dog.

Clad in his spotless Nomex turnout coat and shiny yellow helmet, Red Hall gave Cole a wink and waved for the reporters to gather around him.

Cole stood outside the fringe of the circled media. Red Hall's management of the reporters was inspirational. He showed the cameramen good angles and provided Jennifer Preene and the other reporters with the harrowing details of the fire attack and dog rescue. It was hard to believe this was the same awkward, bumbling fireman he had known for so many years. Illuminated by camera strobes, Red Hall carried the distinctive air of a polished spokesman. As Cole stood admiring his friend, he was once again spun around by the shoulder.

This time it was Captain Decker. Decker stood well within Cole's comfort zone. He wore his helmet cocky and slightly back on his head. The helmet strap hung loose under his chin, a habit of an old soldier, a veteran of too many wars. Decker's turnout coat was also wellseasoned. Scarred with char and stained with soot, his coat showed no clean fabric over the shoulders, the usual sign of air bottle straps, nor did his face show the reddened pressure marks of an air mask. Decker did not believe in using a breathing apparatus. Captain Decker was a *leather lunger*, believing air bottles were for weaklings.

"Ventilation is complete," reported Captain Decker. "We're out of here." Captain Decker's demeanor was curt and abrasive.

The sound of water slapping against the bedroom window casement captured both their attention. A straight stream traced the window frame, boring water into the stubborn recesses of the casing. Occasionally missing the jamb, the water stream shot across the front yard wetting the street and the side of Truck 45.

Shaking his head in disgust, Captain Decker headed for his ladder truck, but stopped for a moment to study the paramedics performing CPR on the dog. Razmunson forced oxygen into the puppy's snout, while Valdez auscultated its chest with a stethoscope. Valdez rocked back on his heels and smiled at his partner. They hooted together and slapped palms over the dog. Razmunson gently set the puppy into the woman's arms and instructed her how to hold the oxygen mask to the dog's muzzle.

Further disgusted, Captain Decker spun around, snarling at Captain Walker. "Looks like we finally found something your crew is good at."

Like a bolt of lightning, from the bedroom, a stream of water hit Captain Decker in the side of the face, flipping his helmet to the side of his head. He clumsily stumbled sideways, struggling to regain his balance. Then, just as quickly as it came, the stream resumed blasting the window casing.

Captain Decker snapped to an erect posture. He glared at the window but was unable to identify the fireman inside who was obscured by steam and water spray. Captain Decker righted his helmet, dumping more water on his head that streamed down his cheeks. He glared at Captain Walker. The side of his face bulged and contorted under the pressure of his torquing jaw. His face was red—most red on the left.

Turning abruptly, Captain Decker stormed to the ladder truck and climbed inside the cab, making no attempt to wipe the water from his face. Undoubtedly, he would steam himself dry.

CHAPTER 12

FALLING ASLEEP

There was more to firefighting than simply introducing water to flame. In actuality, extinguishing fire was the easiest phase of firefighting operations. As the exhausted firefighters were agonizingly familiar, the hard work began after the fire was out.

After the knockdown, when the firefighters were no longer driven by adrenaline, came overhaul. The firemen disassembled and examined everything that could harbor even the smallest ember. Door molding and window casings were torn off. Charred, water-laden mattresses were dragged from the building. Melted, smoldering mounds of polyester clothing were pried from the closet floor and thrown outside. All compromised drywall was pulled from the walls and ceilings until there was no place left for embers to hide, and the fire was declared officially out. Proper overhaul safeguarded the resident from further fire danger, fended the structure from additional damage and protected the district captain from an embarrassing rekindle.

With overhaul complete, Fire Captain Cole Walker declared his first structure fire, "officially extinguished." Barker collected the dirty axes, pike poles and nozzles, while the firemen rolled the dirty hose. Spent from aggressive firefighting, Putnam flopped wearily into the fire engine jump seat. The engine and squad drove away from the darkened building, relocating their work to the fire station.

Back at the firehouse, low pressure air bottles were exchanged with fresh ones. The axes and other tools were cleaned and returned to their shiny brackets. The fire engine was reloaded with clean, dry hose.

In the office, Cole put the finishing touches on the fire report. He set down his pen and massaged the bridge of his nose. He joined the other firemen who had migrated to the dayroom recliners. Within minutes, everybody had lost their struggle with their eyelids - except Valdez. Like the bell to a boxer, Putnam's snore signaled Valdez to begin his scheme.

Lifting his head from the recliner, Valdez studied the other members of the crew. Satisfied that they were sound asleep, he slowly and methodically released the recliner chair's footrest, then stopped at the first squeak fearing any noise could wake the men. Instead, he rolled over the arm of the chair, crouched on the floor and again surveyed the firemen. Bleve raised his head lazily but lowered it in disinterest. Confident that they did not hear him, Valdez snuck out of the dayroom and headed upstairs to the dormitory.

Besides eating, nothing was more sacred at the firehouse than sleeping. And no one was fonder of the activity than Richard Valdez. And no one was more proficient at depriving Valdez of his cherished slumber than Terrence Putnam. It was common knowledge amongst the firemen at Station 61 that Terry Putnam was a blue-ribbon snorer. If graded on an artistic scale, Putnam's snoring was freestyle and impressionistic. He could snore on his back, on his side, and on his stomach. First loud, then soft, then nothing at all. Anyone unfamiliar with Putnam's Cheyne-Stokes might suspect that he died, but to Val's dismay he didn't. Some say Putnam could snore with a mouthful of water, which Valdez disproved during a 3:00 a.m. experiment.

The crew's tactics toward Putnam's snoring eroded to every man for himself. Razmunson usually went to bed with earphones and the latest CD of *Classical* or *Heavy Metal*. Cole went to bed with earplugs in the summer and an extra pillow to cover his head in the winter. Barker just tried to beat Putnam to bed by ten minutes or so, in hopes of falling to sleep before Putnam began his serenade. Richard Valdez was not so lucky. He had the sleep fragility of a soap bubble, while Putnam had the vocal fortitude of the *Bismarck*.

The high gloss enamel paint behind Putnam's headboard betrayed the variety of objects Valdez had hurled during the night. Scuffs and dents on the wall evidenced the shower of boots, high-speed rolls of toilet paper, books, magazines, bars of soap and a starburst stain from a semi-empty bowl of raspberry sherbet. Valdez's attempts to squelch Putnam's nocturnal vocalizing had met with miserable failure. He knew the only alternative left was shock therapy.

Three weeks prior, Valdez took a look in the mirror—a long look. He studied the dark, wrinkled shadows under his eyes. They did not wash off. They did not dissipate after two hours of cooling beneath two frozen steaks. Valdez blamed Putnam for ruining his good looks. But, for Richard Valdez, this was more than his loss of sleep; it was a violation of his person. He did not merely dislike Putnam's snoring—he hated it. He loathed it. He dwelled on it. He wanted to pay the Putz back—not with a gentle shake, but a jolt, or shock, or an explosion; something to make him pay.

For Richard Valdez, the situation looked bleak. His remaining options lay between spending the rest of his career sleeping in a recliner in the dayroom, or lying awake until he could go home to sleep and hopefully not on the freeway before he got there. Then, miraculously last month, during a midnight television program, he was struck with inspiration. Watching the late, late, late-night sports show on ice fishing, he saw a glimpse of a plan. It was a crazy plan. It was a delusional scheme of a demented man under the duress of sleep deprivation. He spent that night plotting and calculating. He couldn't sleep from his own excitement. With any luck, he would cure the Putz with shock therapy.

Valdez snickered every time he thought of his covert operation. He hadn't done anything this devious since he put the paramedic institute's medical cadaver face down in the hospital therapy pool. The ER doctors performed CPR for over two hours on the unidentified drowning victim before one of the interns recognized the corpse. But that paled against his plan for the Putz.

The present time was optimum to unfold his plan. Several elements were conveniently in alignment. Everything he needed was already assembled in the dorm. The dormitory's linoleum floor was freshly waxed. And the crew was extra tired; they'd sleep hard. Tonight, was the night he would initiate *Operation Pole Hole*. Tonight, Richard Valdez was going to get a good night's sleep.

Operation Pole Hole was accomplished in stages. While Putnam slept, Valdez would tie a length of rope around the frame of his bed. The rope, just long enough to stretch to the pole hole, would be tied to a free weight. Valdez would simply drop the weight down the pole hole and after that, physics would take over. The freefalling weight would pull Putnam's bed to the pole hole on the opposite side of the room. The brass pole would stop the bed and jar Putnam from his sleep. A secondary line would be attached to Putnam's bedding. So, as his bed came to a sudden stop against the pole, the remaining inertia of the falling weight would yank the sheets off the sleeping giant. In a single instant, Putnam would be crashed awake and stripped naked.

Valdez left nothing to chance. Earlier in the week, he had tested each bed to determine which one had the smoothest casters. Luckily, it was his own, so switching his bed frame with Putnam's was simple. The most perplexing obstacle was the weight factor. How much weight was needed to pull a bed and nearly 300 pounds of payload across 15 feet of floor? The problem stymied Valdez. There was no practical way to test the maneuver beforehand. His solution to the weight dilemma was simple—the more, the better.

It took Valdez two weeks to inconspicuously disassemble and reassemble the barbells to create a heavy dumbbell that would fit through the pole hole. His clandestine efforts were frequently delayed because firemen on the other shifts would disassemble the goliath dumbbell to reconfigure their own. But today, miraculously, the 300-pound curling bar sat in the corner as inconspicuous as a train axle. Everything was as close to perfect as it would ever be. Tonight had to be the night.

To the rest of the crew, the evening appeared to be no different than any other. Putnam stayed up late, giving everybody a chance to reach their REM sleep before he rocked them out of it. He said, "good night" to Valdez who was slumped in a recliner. After he left, Valdez lowered the television volume and waited for Putnam to give him the signal to come upstairs.

Kathryn Donaldson expected insomnia again tonight. She wished her boss would make his promotion decision soon. She couldn't stand the tension. She couldn't take the stress another week.

She slipped into the bathtub. The warm water caressed her skin like a massage.

The warm bubble bath was nice, but a promotion to European Sales Rep would be even better. And the new job would come with a nice change of scenery. She closed her eyes and slid deeper under the bubbles and allowed her thoughts to roam the provincial towns of Europe.

After her bath, Kathryn Donaldson sauntered to the bedroom. The sheets felt cool against her warm, damp skin. Closing her eyes and shivering away the cold, she thought of white fluffy sheep, hundreds of them, all jumping over a white rail fence back dropped against green rolling farmlands somewhere in the European countryside.

Valdez paid no attention to the muted television; his mind was upstairs in the dorm. He reviewed his plan, tracing and routing the rope connections. He shut off the television and sat upright in the recliner, straining, listening for the signal. To the inexperienced ear, what would have sounded like a heavy wooden chair sliding back and forth across a hardwood floor was, in reality, Putnam's snore; his signal for Valdez to start.

Valdez tiptoed upstairs and gingerly pushed open the dormitory door, allowing a thin crack of light to slice into the lightless room. Slipping inside, he stood motionless, waiting for his eyes to adjust to the dark. Blinking hard against the softening blackness, he studied each bed. Captain Walker and Barker, with cotton plugs in their ears, slept motionless in their bunks. Razmunson slept face down with a pillow pulled tight over his head. The next empty bunk was his own. Next to it, the last bunk on the end lay the Putz, on his back, snorting and gagging as usual.

Sliding his stocking feet across the freshly waxed floor, Valdez eased his way to the exercise area at the far end of the dormitory. He pulled two sections of nylon cord from behind the modified dumbbell. He tied one cord to the bed, centering the knot between Putnam's monstrous bare feet jutting well beyond the end of the mattress. His soles glowed white in the darkness, and they smelled of cool perspiration. Annoyed by the smell, Valdez pulled his tee shirt over his nose like a bandit and continued tying the rope. He tied the secondary cord to the bed sheets, then to the main rope. Putnam stopped breathing, freezing Valdez in an incriminating, mid-knot position. After what seemed like an eternity, Putnam resumed his snoring.

Thawed from his scare, Valdez shuffled back to the 300-pound dumbbell, paying the cord from its coil. Humoring himself, he noted the irony of a 300- pound dumbbell at each end of the rope. He rolled the dumbbell to the edge of the pole hole. The unevenness of the cast iron weights made the dumbbell teeter precariously near the edge of the hole, then it slipped from his control. He grabbed at the dumbbell, straining to stop it from falling into the apparatus room below. As he rolled it back from the hole, the weight continued toward him until it reached its center of gravity and rocked to a stop.

He tied the cord from Putnam's bed to the barbell, leaving enough slack so the rope wouldn't stretch taunt until the weight was in free fall. He then tied a second, smaller line that he called the primer cord. He would pull on the primer cord to get the weight rolling. He led the primer cord around the brass pole, then back to his bunk so he could trigger his creation from his bedside.

Valdez slid under the covers and held his breath as he pulled on the primer cord. The weight did not move. He had failed to consider that the primer cord was tied to center of the bar, giving it no leverage to make the weight roll. He jerked on the cord—

nothing. He pulled off his socks, put down his bare feet to better grip the waxed floor. Taking up the slack in the primer cord, he wrapped the excess line around his fist for a better grip. He jerked the cord, causing it to make a delicate *ting* as it slapped against the brass pole. The weight rocked.

He arched his back, straining against the cord as it cinched deep into the soft flesh of his hand. The weight gave way. The primer cord pulled easily. Val lay back on his bunk, pulling the cord and listening to the iron weights grind across the floor. Suddenly, there was a brief instant of silence, followed by the nylon cord painfully ripped from his grasp.

When the 300-pound weight began its free fall, the foot of Putnam's bed jerked violently to the left aligning itself with the hole as the nylon cord sizzled over the corner edge. Putnam could only raise his head against the inertia as his bed rocketed across the room and slammed into the brass pole with a loud *clang*.

The head of the bed shot into the air as the foot casters fell into the pole hole. Putnam's body slid down the mattress, disappearing under the covers and jettisoning from the bottom of the sheets like a burial at sea. Putnam's cadaverous white body flashed from under the sheets, his forehead struck the pole with a deep knell. Perhaps out of instinct or pure luck, Putnam latched onto the pole and slowed his descent. The bed teetered in the air until the second rope snapped at the sheets sucking them down the hole. The bed fell back, slammed to the floor, bouncing the casters out of the hole and rolled backwards into the shadows. Below, on the apparatus floor, Putnam sat dazed, straddling the pole, rubbing his forehead and staring dumfounded at the dumbbell rocking beside him. He pulled the crumpled bedding from his head and gazed at the empty hole above.

Although the event was an application of physics, Putnam perceived it as mystical. Underneath the cacophonous discourse of his own snore, he was deep within the tranquil recesses of slumber, floating on a calm lake with the beautiful Evelynn Dewitt. Her blond hair glowed iridescent in the bright sunlight. White swans glided past their boat as Putnam, the ship's captain, sailed his guest through paradise. A light breeze played with the sail. Putnam stood poised at the helm as the lovely Miss Evelynn dipped her finger in the lake, making small ripples in the lazily passing water. Suddenly, unannounced, a black, sinister boat commanded by Pirate Valdez appeared from the starboard side. Dressed in buccaneer clothing, Valdez sided their vessel and kicked open his treasure chest overflowing with chocolate chip cookies. Bowing gracefully and accepting his gloved pirate hand, Miss Evelynn changed ships in the middle of the lake. Valdez bellowed a swashbuckler laugh. With a mighty shove of his black pirate boot, he separated the two boats. Putnam sat motionless, frozen at the helm. He tried to move but couldn't.

As a parting gesture, Pirate Valdez blew into Putnam's sail. So powerful was the gust that it knocked Putnam flat against the transom. He could do nothing as the boat careened toward shore. Struggling against the boat's inertia, Putnam was only able to lift his head slightly, just in time to witness his ship on a collision course with a tribe of angry horse people. The ship sped faster and faster and crashed into the horde of vicious horse people. The impact threw Putnam from the stern directly into the main mast and, clutching it for dear life, he slid down and down until he reached the bottom of the apparatus floor.

Sitting lethargic at the base of the brass pole, Putnam looked up and all around, then back up again. He rose slowly to his feet and tried to separate reality from imagination. Dazed and confused, he rubbed at the bump on his forehead. Instead of figuring out his present situation, he collected his pillow and blanket and headed for the recliners, hoping it was not too late to rescue his fair maiden.

Dennis Upman made his last delivery in the industrial district on the outskirts of Seattle. The tractor was easy to drive now that the two trailers were offloaded. He was glad to get off the road. He wasn't due in Torrance, California until the end of the week. It had been a long two weeks with no sleep.

The motel manager was suspicious when the skinny, unshaven, grungy trucker paid for three days in advance–in cash. Upman's only request was not to be disturbed. The manager hesitated at the dirty trucker's request.

Upman flashed his union card. Partially convinced, the manager relented and slid the room key to the tired trucker.

The room was at the rear of the motel. It seemed to take forever to walk there as each step drained him of energy. Dragging from exhaustion, Dennis Upman struggled with the dirty doorknob, the keyhole blurred. Finally, open, the musty room wreaked of stale cigarettes. He fumbled the plastic *Do Not Disturb* sign over the doorknob and flopped onto the bed. The sheets felt soft and cool. He kicked off his shoes without untying them, then fell from consciousness and slept for the next three days.

CHAPTER 13

DISCIPLINE

As usual, Val's lunch was a success. The homemade salsa, beans, carne asada, authentic Spanish rice and fresh tortilla chips weighed like a rock in Cole's stomach, making his skin feel flushed and warm. He struggled with his heavy eyelids as he wrote in the station journal garbling his words and scratching streaks of ink across the page each time he nodded asleep.

The phone rang. Cole quickly picked it up, then snapped wideawake the instant he heard the Chief's voice. They exchanged insincere pleasantries.

Chief Pierce seemed to be in an unusually pleasant mood, complementing Cole on his good job with the residential structure fire last shift. The Chief was pleased with the news coverage of the dog resuscitation.

"Nice touch," said Chief Pierce, oddly chipper. "We sure impressed the animal people," referring to the animal rights activists. Chief Pierce said he was still receiving faxes from the animal people wanting interviews, offering donations and suggesting the fire department teach animal CPR classes to senior citizens. Then it came. Cole recognized the Chief's mood swing quick enough to jerk the phone from his ear.

"I also received a few other faxes," said the Chief, his anger evident in the volume of his voice. "Like the fax from Mayor Jenkins. He seems to pride himself in providing as much detail as possible on your parade demeanor. In addition, he included a repair bill for three tubas, two trombones, and one bass drum!" The Chief yelled so loud that his voice cracked at its crescendo.

"I also received a fax from the Palomino Pals in Duarte, whoever the hell they are. They want me to pay for thirty-six hours of horse therapy." Chief Pierce paused to allow his words to steep. "What the hell is *horse therapy?*" he shouted, sounding like the fire chief Cole was familiar with.

Cole did not immediately answer, knowing he was not supposed to. He allowed the Chief to vent and blather on while he contemplated horse therapy. Cole envisioned a horse lying on a leather couch while a bearded psychiatrist in a turtleneck sweater took notes. He was startled back to the present conversation by Chief Pierce's booming voice.

"I've got to hand it to you, Captain Walker. You are resourceful. Wherever you go, you provide your own emergencies!"

"Here comes the train incident again," mumbled Cole to Captain Hastings' portrait.

"I thought after the train incident, we were clear on the mission of the fire department, but apparently not!" screamed the Chief. "Our goal at the fire department, Captain Walker, is to stop emergencies, not create them!"

Cole got comfortable, placing his elbow on the desk and resting his chin in the palm of his hand. He held the phone at arm's length and could hear Chief Pierce loud and clear.

"Let me establish some guidelines for any further incidents you may come across, Captain Walker. You may want to write these down. Ready?"

"Ready."

"Do not run over any more horses. Do not drive your fire engine into crowds of people. Do not run over little old ladies crossing the street. Do not set the hospital on fire. Do not down aircraft. Do not sink any ships. Do not stomp the skulls of baby kittens on the steps of City Hall. Do not drive your fire engine off a pier unless you can guarantee me you are strapped to it! Do I make myself clear, Captain Walker?"

"Yes, sir."

"Did I leave anything out, Captain Walker?"

"No, sir."

"Then don't screw up again unless you want to spend the rest of your career serving soup in the headquarters cafeteria!"

"Understood," said Cole to the dial tone. He understood all too well. He knew the Brookfield annexation would fail. He knew Mayor Jenkins would fight him on every issue. He knew at the end of the contract trial year, Brookfield would revert back to a private fire department, and he'd go back to headquarters—probably to serve soup.

Cole Walker stood up from his desk and shrugged his shoulders at his reflection in the portrait of Captain Hastings. "I only have a couple of weeks left as a Captain, so I think I'm going to enjoy it." It was time to drill the crew.

The end of Firefighter Recruit Terrence Putnam's probation was at hand. Putnam would be a boot for three more shifts, so the crew conspired to celebrate his passage early. Generally, recruits expect hazing on their last day of probation, but taking down a man the size of Putnam required the element of surprise.

As part of the preparation, Cole had carefully scheduled their EMT drills. Basic first aid was the overall theme. Earlier in the week, the subject was bandaging. Last shift, it was splinting. Today, they would drill on backboards and patient immobilization.

Assembling on the open space of the apparatus floor, the crew gathered around Captain Walker and the Miller board. The Miller board was a backboard designed for total body restraint. The lower end was split for lashing each leg individually and the top of the board narrowed at the shoulders for stabilization of the head. There were additional straps for the patient's chest and waist. The Miller board was specifically designed for immobilizing victims of spinal fractures, or for restricting the movements of hostile patients – of which Putnam was soon to be.

As with his previous drills, Captain Walker explained the procedures for the equipment use. He emphasized that strapping the patient tightly to the backboard was essential so he could be transported completely immobilized or, in case of nausea, safely flipped on his side without jeopardizing his spinal alignment. At the conclusion of his instruction, Captain Walker laid the board on the cement floor and made his request. "I need a volunteer."

All eyes turned to Putnam. His rank as boot, coupled with his eagerness to participate, overshadowed any sense of suspicion he should have felt. Since his probation was not officially over for another three shifts, his guard was down. Unwary, he straddled the Miller board then lay down on it. The six-foot Miller board was about six inches too short for Putnam. His large boots dangled off the foot end of the backboard.

Next, several two-inch wide nylon straps were run across his chest, through slots on the side of the Miller board, then doubled back onto themselves with Velcro. Then the hip strap, two thigh straps, and four leg straps were pulled tight. The head strap was last. Up to this point, Putnam had his head raised watching his own imprisonment.

The head harness was a single unit, consisting of a padded head *bucket* with forehead and chin straps. The bucket was tightly strapped over Putnam's pumpkin-sized head and secured to the board. At the conclusion of Putnam's immobilization, Cole pulled on each of the ten straps checking their tightness. Some, he retightened. The girth of the large fireman's barrel chest did not leave much strap to overlap against itself. The two-inch nylon straps were miniaturized against his massive body—like threads across Gulliver.

"Okay, Terry," said Cole, "Try to get out."

Putnam reached with his unrestrained hands and released the Velcro straps.

"No, not that way," said Cole. "Try to bust free."

Putnam squirmed and grunted. He lifted his massive neck against the head straps, then arched his back against the board. The Velcro crackled but held. He relaxed. His neck veins receded, but the redness remained in his face.

"I can't do it, Cap," he reported.

"Good job," Cole complimented his crew.

Cole continued his instruction, especially to the paramedics. He pointed out that the torso straps did not encompass the victim's arms or wrists as Putnam had demonstrated. In the case of a combative patient, free arm movement was dangerous and detrimental to further treatment such as bandaging or establishment of an IV line. Wrist restraints should be used. Razmunson pulled two leather restraints from the paramedic truck. The wristlets were made of hard leather with a cushioned interior lining. They were three inches wide with size adjustment slots. Once the wristlet was wrapped around the patient's wrist, a secondary leather belt was used to lock it and secure the wristlet to the backboard.

Fitting the leather restraint to Putnam's wide wrist was a tight fit using the largest adjustment hole. The secondary strap was run through the metal eyelet and tied to the backboard, effectively securing his left arm at his side. Securing his right arm proved much more challenging.

Up to this point, Putnam was cooperative and complacent, but when they attempted to strap the leather wristlet to his right wrist, he realized that the drill had expanded beyond the instructional phase. He twisted his arm out of the wristlet just before Valdez latched the buckle. He raised his massive arm and reached for the Velcro straps across his chest. Valdez leaped on Putnam's free arm but was easily lifted into the air. Razmunson came to Valdez's aid by pulling back on Putnam's thick arm, but he could only hold it in check.

Putnam's arm wavered unyielding. Both Valdez and Razmunson pushed with all their body weight to force the big man's forearm to his side. Putnam's groping hand found the loose leather wristlet. He gripped it tight. While Valdez and Razmunson wrestled with Putnam's right arm, Barker struggled to remove the wristlet from Putnam's clenched fist. Pulling back on his sausage-sized fingers, one-by-one, Barker worked the restraint free from Putnam's grasp.

Putnam violently heaved his chest up and arched his back, straining against the Velcro straps. Cole sat on his chest. All of the Velcro crackled, Putnam's left leg broke free. He lifted his giant leg into the air, gaining leverage as he heaved his chest again and arched his back hard. Cole rose up and down with each breath of the seething giant. Putnam's breath blew hot against Cole's ear.

"I don't suppose you guys could hurry up," said Cole, riding Putnam's heaving chest and pressing his hands against the crackling straps.

As Barker struggled to get the wristlet around Putnam's wrist, he spoke through his clenched teeth, "We're... almost... there..."

With his free leg, Putnam pushed his heel against the floor, spinning the backboard in a circle. Valdez and Razmunson lost their balance and fell backwards, allowing Putnam's massive arm to rise free. It was locked in the wristlet, but not yet tied to the backboard. They dove on Putnam's arm, forcing it back down to his side.

"Hurry up, you old fart," Razmunson yelled at Barker as he fumbled with the leather straps. Finally, tying the straps to the backboard with several granny knots, Barker jumped up and strutted like a rodeo calf wrestler. The four firemen, exhausted and breathing heavy, stood on the periphery admiring their work as Putnam used his free leg to push himself in circles.

The strapping of Putnam's left leg was anticlimactic as he offered little resistance to the inevitable. Once tightly secured on the board, the four firemen groaned as they lifted the big man and carried him outside to the hose tower.

Hoisting Putnam up the hose tower was another difficult feat. Barker grabbed the rope that looped through the upper pulley and tied it to the top handles of the Miller board. Valdez and Razmunson hung on the other end of the rope. Putnam did not budge. Barker added his weight to the effort until Putnam was upright and teetering on his toes.

"Don't drop him!" shouted Cole, running over to add his weight to the rope. He envisioned the large fireman strapped helplessly to the backboard falling forward onto his face. In unison, the four firemen hung on the rope and hoisted their captive six inches off the ground. Barker secured the rope.

"Valdez," commanded Captain Walker in a voice distinctive of a military commandant. "Retrieve the condiments."

"Aye, Aye," said Valdez, mocking a salute. Valdez and Razmunson trotted off to the kitchen. Shortly after, Valdez returned with a cardboard box filled with kitchen condiments.

Swaying from the hose tower, head strapped back, Putnam strained his eyes downward to see inside the box. Razmunson set beside the condiments a kitchen drawer full of, in his words, "application tools."

"Recruit Terrence Putnam," said Captain Walker, using his best Captain Bleigh impersonation. "For the infraction of successfully completing your probationary period, you have been sentenced to condimentation." Cole spun a sloppy about-face and addressed the crew. "Commence with the condiments."

The honor of the first application was given to Engineer Barker for his seniority. He promptly obliged by cracking an egg on Putnam's head. Barker gently dislodged the yolk from the shell allowing it to sit on the big man's head. What followed next was a free-for-all. Valdez sprayed Putnam with a plastic squeeze bottle of mustard, striping his uniform yellow. He stepped aside allowing Razmunson to fling spoonfuls of mayonnaise at the swaying recruit. The mayonnaise impacted with a *fwap*, splattering little white droplets in every direction. Next, Putnam was basted with relish, then ketchup, tomato paste, and powdered sugar. Putnam asked for a taste of powdered sugar. Razmunson dumped most of the box on his extended tongue. Cornstarch, marmalade and last shift's sour kraut were poured direction into Putnam's pants. Their hazing was interrupted by the sound of a squeaking fan belt.

A beige station wagon pulled into the parking lot. The driver's face was hidden by glare on the windshield. They all stood, or hung, in silence staring at the car. The driver's door opened, and the chassis rebounded as the gargantuan occupant stepped from the car.

"Mrs. Broadenhoff," said Cole, as he briskly walked to meet her. He stopped directly in front of her, hoping to block her view of Putnam dangling from the hose tower. But, since she was at least six-inches taller than Cole, his effort was in vain.

"Hell-o, Captain Walker," said Helga Broadenhoff, staring at Putnam. "Vas is going on here?" she inquired, fixated on Putnam swaying from a rope.

"Hell-o, Mrs. Broadenhoff," said Putnam, in his usual cheery voice. He smiled at her, peering through one squinted eye, the other eclipsed with a pickle slice. The top of his head was mounded with spaghetti.

Helga Broadenhoff stood silent, switching her puzzled stare from Putnam to Captain Walker, Captain Walker to Putnam. She forgot about the three- quarters of a chocolate cake she held in her hands. She had rescued it from the Brookfield Lady's Bowler's Auxiliary party.

Cole reached for the cake. Helga Broadenhoff held tight to the plate, mesmerized by the colorful, dripping fireman swaying from the hose tower. Cole gave the plate a tug, jerking it from her grasp.

"Oh, is this for us?" he asked as her grip snapped from the plate. Speaking louder, he said, "It sure looks good. Did you make it?"

"Yah…" she said drowsily. She lowered her stare from Putnam to Captain Walker. She stared blankly, her lips slightly parted.

"Did you make the cake?" Cole repeated, raising the cake in front of her face hoping to break her hypnotic trance. "Well, thank you very much, Helga." He grabbed her elbow and led her back to her car. "You don't mind if I call you Helga, do you?"

"Yah…Yah…" she replied. She shuffled beside him, staring back at the colorful fireman hanging from the hose tower.

"Good-bye, Mrs. Broadenhoff," said Putnam, waving with his fingers, his wrists still strapped at his side. He blew a multicolored spray, as the condiments dripped from his upper lip.

"Vas is going on?" asked Helga Broadenhoff, as she squeezed into the front seat, dropping her car lower.

"Well," whispered Cole confidentially, "Disciplinary action, ma'am. It's best to keep this secret, Mrs. Broadenhoff. Excuse me... Helga. Can you keep this a secret, Helga?"

"Yah?" she said, wide eyed, pondering his request. "Oh, yah. Secret, yah, yah," she said, then fumbled her car keys into the ignition. She started her car and slowly backed down the driveway. Leaning over her steering wheel, she watched Putnam until he disappeared behind the corner of the fire station.

Cole watched her as she slowly backed down the driveway, then he turned his attention to Putnam. As he walked back to the hose tower, he stopped at the trash dumpster and raised the cake. "Anybody hungry for chocolate cake?" The answers were three thumbs down and one "yuck." He lifted the lid to the dumpster but paused and dropped the lid. Instead, he pressed the cake into Putnam's chest.

Captain Walker stepped back and stood directly in front of Putnam, just beyond his spray of condiments. He turned to Razmunson and ordered a bag of flour. Cradling the bag of flour in one arm, Cole kicked the bottom corner of the backboard sending Putnam into a slow spin.

"Firefighter Terrence Putnam. With the powder vested in me as the captain of this here fire station, I hereby declare your rank and title changed." He tore open the bag and, with each revolution of the backboard, threw a fistful of flour at Putnam, momentarily obscuring him in a cloud of white dust.

"In the name of the Fire Chief..." Flour.

"...and the Mayor of Brookfield..." Flour.

"...and the Los Angeles Fire Department, I pronounce you Firefighter, permanent and stationary."

Flour. Flour.

"Valdez!" barked Captain Walker, clapping the flour from his hands. "Clean off this firefighter."

After saluting the captain, Valdez pulled a length of fire hose that was preconnected to the yard hydrant. He pointed the nozzle at Putnam and pulled back the bail handle. The powerful water stream struck the edge of the backboard sending Putnam into a spin. As the water spun him faster and faster, colorful fans of red, yellow and orange marmalade sprayed from his body as he reached blender speed.

CHAPTER 14

"EAT ME"

Putnam delicately held the two theater tickets in his hand like flowers. He reassured himself that *CATS* was a good choice for a first date because Evelynn likes animals and there must be animals in *CATS*. He practiced his lines again, speaking to the tickets, praying he would talk without stammering. "Evelynn, would *you* go out with me? Evelynn, would you *like* to go out with me...please?" He shook his head and corrected his sentence, "Would you *please* go out with me?" He cleared his throat. "Evelynn, would you please come with me to the show? It's about animals."

Nothing sounded smooth. He wasn't very good at this sort of thing, but he was determined to do it – today.

He put the tickets in his uniform pocket and gave his pocket a reassuring pat. He rehearsed his lines to his open locker, mumbled his lines while brushing his teeth and whispered them while he gathered rags and brass polish. He and Valdez had been assigned to pole duty. But, by the time the job was done, he had done most the work.

Putnam picked up the armload of blackened rags, while Valdez combed his hair in the reflection of the polished brass. Putnam had polished the high parts and the low parts, while Valdez supplied the entertainment. He talked about girls; the ones he dated, the ones he dumped; the dancer, the gymnast, and the twins.

"Let's change the subject," said Putnam.

"To what?" scoffed Valdez. "Tractor rebuilding?"

"I don't care to hear about all the people you screw around with," said Putnam.

"Well, it's better than sitting around watching your overalls fade," said Valdez defensively. "Speaking of wasting time, are you going to do anything with the Cookie Girl? Or should I? Besides, I don't think she's your type anyway."

Putnam had had it with Valdez. He stepped closer until the two men were chest to chin. He aimed his thick index finger at Valdez when the doorbell rang. The interruption came at a good time. Putnam wrapped his huge arms around the brass pole and slid from sight.

When Putnam saw Evelynn through the window of the front door he was too stunned to smile. This was the first time they had seen each other since the chocolate cake incident. They exchanged smiles through the door glass as he spun, and spun the door handle, his hands slippery with brass polish.

Framed in the oak doorway, Evelynn beamed radiant in the sunlight. Her vanilla hair glowed like a halo anointing her head. She smiled and fussed with the striped bow on the cookies. Putnam offered a handshake, but quickly withdrew his hand when he saw it caked with polish. He hid his hands behind his back, pressed against the doorjamb and motioned Evelynn to enter. As she passed, Putnam admired the open back of her floral sundress. Spaghetti straps crossed over her lightly freckled shoulders and, with each step the curled ends of her hair danced across the back of her neck.

She spun capriciously and extended the platter of cookies. She nervously bit on her lower lip. "I brought a peace offering," she said, lifting the plate of cookies under his nose. The sweet aroma of chocolate chips and rose petals lofted into Putnam's flaring nostrils. He feigned interest in the cookies while ostensibly admiring the Cookie Girl. With fleeting glances, he studied her riant eyes, the innocent way she shifted her weight from foot to foot and the light freckles that spilled across her shoulders before disappearing into the shadow of her cleavage.

Attention to the Cookie Girl was not lost to just Putnam. When she entered the kitchen, the room erupted with the sound of sliding chairs as the other firemen stood and offered their seats to the lady. She accepted Barker's.

Gently sliding his chair beneath the Cookie Girl, Barker complimented her on her fragrance, saying she "stinks pretty" and openly wishing his wife would wear some perfume rather than the sour milk she normally slaps on her neck.

"If it keeps you at a distance, I understand her methods," said Cole, retaking his seat at the head of the table next to the Cookie Girl. Putnam headed for the chair next to Evelynn, but Razmunson slid into it first. Putnam moved for the seat across from her, but Valdez beat him to it. Putnam took the farthest seat away and slid the plate of cookies to the center of the table in an effort to break up the gang-stare at Evelynn.

Acting unusually playful, Valdez spun the plate of cookies between his hands and interrogated the Cookie Girl with a thick Gestapo accent. "How do vee know these cookies are safe to eat?" he asked, crushing his large eyebrows and squinting at the fraulein. He spoke again through tight lips. "You poison us once, yah?"

The Cookie Girl gently removed the striped bow and stuck it to the back of Valdez's hairy hand. She lifted the corner of the cellophane, selected a small cookie and raised it to her mouth. Valdez quickly grabbed her hand and guided it back to the mound of cookies.

"Not zat von," he said, raising one eyebrow. "Eat zis von," he said, picking a cookie from the bottom edge of the stack. He continued to hold her left hand as he leaned across the table and held the cookie to her lips. As Evelynn's mouth opened to receive the cookie, she suddenly pulled Valdez's hand and took a large bite nearly nipping his fingers. Putnam couldn't take his eyes off their hands. Valdez was holding both of her hands. As Valdez ate the remainder of the bitten cookie, everybody else pulled cookies from the bottom of the stack, except Putnam. He lost his appetite and stormed from the kitchen.

He stood in the emptiness of the apparatus room and pulled the *CATS* tickets from his pocket. "I guess I'm not her type," he said as he tore the tickets in half. He threw the pieces at the trashcan but missed. "I can't do anything right," he said and sulked away.

For the remainder of the shift, Putnam could not shake the vision of Evelynn hand-in-hand with Valdez. "Valdez of all people," mumbled Putnam, as he clanked plates of leftovers back and forth in the refrigerator. Empty handed, he slammed the fridge door. The clatter of the condiments did little to arouse the other firemen who sat in the recliners, eating popcorn and watching the evening movie.

"Hey, Putz…"

The voice cut into Putnam's spine like a chilled stiletto. Valdez eased back in his recliner, and waving his empty bowl above his head, not taking his eyes off the television, "Get me some more popcorn."

"Get it yourself," snapped Putnam, shoving through the kitchen door. "I'm off probation." He stomped into the apparatus room, not sure of where he was going.

"What's his problem?" asked Razmunson.

Valdez just shrugged his shoulders, though he did know the reason for Putnam's distemper. He decided it was time to twist the knife.

The sound of mail scraping through Kathryn Donaldson's mail slot nearly made her run to the door. She recognized one envelope among the mess of letters strewn across her front entry; it had the company's logo. She picked up the envelope and spun it pensively, then tore it open. She read and reread the letter. It said that her promotion application was accepted, and she was invited to the chairman's personal interview. She danced to the kitchen and popped the cork on the champagne.

Recent events at the firehouse deteriorated the relationship between Valdez and Putnam. The standoff between the two firemen continued to escalate. Small incidents festered and swelled out of proportion, layering one upon another like blankets over hot coals. Trite instances, unnoticeable to anyone else, became major irritations between the two men. The reason for each man's anger was obscure and subjective. For Terry Putnam, he saw Richard Valdez as a sly, manipulative, troglodytic man bent on stealing the Cookie Girl. Yet, in the deepest recesses of his rational mind, he knew that he himself was to blame. His timidity, his inaction, and his fear of rejection all surfaced in the face of Richard Valdez.

Richard Valdez, on the other hand, hated weak people. He deplored Putnam. He hated Putnam's farm boy posture, his laggard speech, his plodding gait, his naivety, and his height. He had a deep, deep resentment for Putnam's size. Richard Valdez probably didn't consciously know it, but he hated all tall people and Putnam was their poster boy. The loss of power, the loss of stature, and the loss of control over Putnam ate at Valdez. Allowing Putnam the final, defiant word last shift was belittling. Having Putnam's defiance in full view of the crew made Valdez's small stature seem even smaller. Saving face became as important as life itself.

The next C-shift, after checking out the equipment on the fire engine, Putnam headed for the kitchen to indulge in the leftover orange juice and muffins he spotted in the refrigerator earlier that morning. Richard Valdez was already in the kitchen.

Already on edge from some sort of dating mishap last night, he braced himself against the counter, waiting for the coffee to stop brewing and the aspirins to start working. Impatient, Valdez pulled the coffee pot from the brewer, while it still dripped. He steadied himself with one hand on the counter, while pouring coffee with his other shaking hand. He hoped the caffeine would quell his throbbing head.

Immediately sipping from the overfilled mug, he lurched forward and spit the coffee into the sink. "Hot" was all he could blurt. The coffee burned the inside of his mouth and the top of his tongue making his head pound even harder. Replacing what had spilled, Valdez refilled his coffee mug to the rim, knowing he would need all the caffeine he could get. Protecting his coffee from spilling and his head from throbbing, he shuffled cautiously to the kitchen's swinging door. With his free hand, he eased the door open about two inches before it reversed directions and knocked him backwards, spilling scalding coffee on his wrist and uniform.

As Putnam stepped through the door, Valdez slapped at the droplets of burning coffee on his pants. Disgusted, Valdez straightened up and shook the coffee from his hand. "Geez, Putz. Why don't you look first before barging through the door?"

"There's a window on your side, too," returned Putnam, not offering an apology. Showing no compunction, Putnam stepped over the puddle of coffee and disappeared behind the refrigerator door. Valdez glared angrily at his half-empty mug and shook his head. He shoved open the kitchen door and stomped out hotter than his coffee.

Waiting for the sound of the kitchen door to stop slapping, Putnam emerged from the refrigerator. He slipped two English muffins into the toaster and poured a large tumbler of orange juice, then drank the remainder of the juice from the carton before tossing it into the trashcan. Outwardly, his actions appeared cavalier, but his thoughts boiled with Valdez—arrogant, domineering, pompous, overbearing Valdez.

The *pop* of the toaster broke his trance. He stabbed his knife into the tub of margarine and forced the cold, yellow wad into the crevices of the muffin. He ate the first half, while buttering the second. He stuffed the remainder of the first muffin into his mouth, grabbed the other muffin and the tumbler of juice, and backed through the kitchen door.

The fire engine and paramedic truck were parked on the front apron, while Barker swept the apparatus floor. Valdez was nowhere in sight, which was just fine with Putnam. He tore a piece of muffin with his teeth and called to Barker, speaking around the wad of bread. "What can I help you with?" Putnam asked, watching Rus as he continued walking around the rear of the paramedic truck.

"Check the BA pressures," said Barker, not bothering to look up from his push broom.

"Will do," answered Putnam, about the same instant he kicked the wash bucket sitting next to the paramedic truck. The galvanized bucket skidded along the wet cement, stopping hard against Valdez, who was crouched, washing the back tire. The bucket stopped suddenly but the water did not. A wave of dirty, soapy water crested over the bucket rim, dousing Valdez. He whipped around already knowing he would see Putnam.

"That does it!" he yelled.

Caught off guard, Putnam reverted back to a preconditioned response. He stumbled backward. Valdez smelled fear and in one fluid movement grabbed the side handles of the bucket and headed for Putnam. Terry broke and ran.

Instinctively running backwards, Putnam retreated into the apparatus room. Barker stood motionless and pulled his broom up tight as the big man brushed past him in full gallop. The rear bay door was closed. Putnam's eyes darted to every door off the apparatus room; all were closed. Any pause to open a door would shorten the precious distance between him and Valdez who was close behind. Lighter and faster, Valdez, carrying a bucket in his hands was closing the gap. Breathing audibly in rapid gasps, Putnam panicked and headed up the open stairway. Valdez was only a few steps behind.

Pulling on the handrail and bounding two steps at a time, Putnam widened the gap. He could hear quick, short shuffling steps following up the stairs. Onto the top floor he leaped. His mind raced. Ahead, the locker room—*dead end*. The bathroom—*dead end*. He slid on the slick linoleum and grabbed at the dormitory's doorjamb. Valdez was on the top step. Putnam shoved open the dorm door, smearing butter and orange juice on the center panel. Once in the dormitory, the pole hole was his only escape. His plan: slide down the pole hole, run across the apparatus room into the freedom of the outside where there was more room to maneuver. But Valdez was too close. Putnam couldn't slide down the pole without Valdez dumping the bucket of water on his head. On his head would be the worst way to get wet.

"I've got you now, Putz. You're mine," said Valdez, in short, gasping breaths.

Putnam was halfway down the dorm and frantic with panic. He was trapped. Out of sheer animal desperation, he jumped to the right and stood on top of Valdez's bed. He snatched up Valdez's pillow with his muffin hand and held it against his chest as a shield.

Valdez stopped at the edge of his bunk, the tempest of dirty water sloshed in the pail. He took a series of sidesteps, back and forth like a broken toy, not knowing what to do. Putnam was standing on Valdez's new comforter, his *dry clean only* comforter. And Putnam was clutching his clean pillow with his greasy hands and precariously waving a glass of orange juice over his new bedding.

Putnam braced himself for the tidal wave that never came. He slowly lowered the pillow from his face as his dim light of realization brightened. Valdez could not throw the water. Putnam realized he was in a safe position. *Sanctuary! Sweet, sweet, sanctuary!* Putnam did a little dance. Stomping his big feet on the plaid bedspread, he spun and tromped in a circle like he was performing some sort of clumsy rain dance. Sipping merrily from his big tumbler of orange juice, he trounced the bedding knowing full well he was agonizing Valdez who stood bedside, impotent with his bucket of dirty water.

As Putnam twirled, bouncing and humming his little victory chant, he studied Valdez with each revolution. For some reason, perhaps an uncontrollable twitch from deep inside his psyche or the response from a crazed man with unimpeachable power, on his third or fourth spin Putnam threw his orange juice in Valdez's face.

The orange juice splattered hard against Valdez's cheeks. Rivulets of orange juice ran across the furrows of his brow and dripped from his nose. Orange droplets beaded on his shoulders and on top of his heavily hair sprayed hair. Streams of orange juice ran down his shirt, down his pants and puddled on the floor. Pulp clung to his reddened face. He visibly shook, but his hands were too full to wipe his face. Through clenched teeth, Valdez bellowed an agonizing roar, as he stood helpless in a puddle of orange juice. He rearranged his grip on the bucket handles so he could point his index finger at Putnam. He shook when he spoke, but his words were perfectly clear.

"This isn't over, Putz. Sometime, somewhere, I'll get you." He held his stare enforcing his sincerity. Putnam's euphoria was chilled by the inflection of Valdez's voice. "It may not be today. It may not be next shift. But mark my words, Putz, I will get you."

Leaving a trail of orange pulp, Valdez sloshed out of the dorm with his pail of dirty water.

Putnam sat on the bed for a long while, long after he heard the shower stop and long after Valdez's steps disappeared down the stairs. He sat. He sat and stared at his empty tumbler and the puddle of orange juice on the floor.

For the next several shifts, Valdez was friendly—very friendly; too friendly. He poured coffee for Putnam and called him Terrence. Valdez opened doors for him, but at every opportunity, whispering in a quiet, disciplined voice, he reminded Putnam, "Today might be the day."

For Terry Putnam, the succession of warning and threats were effective at first, but later reversed their affect. Early on, he stayed awake late at night worrying before falling into a tumultuous sleep. As days passed and the number of unfulfilled threats accumulated, their affects attenuated. Putnam grew complacent, no longer looking behind doors and around corners. He figured that living in constant fear could be the payback itself. As time passed, he grew defiant.

With each unfulfilled prophecy of retribution, Putnam grew less and less concerned. And he said so. Each of his comments became more and more brazen.

"Are you prepared, Putz?"

"Take your best shot."

Two weeks of idle warnings deflated Valdez's foreboding and inflamed his ego. His last statement, 'Are you prepared?' was intended to strike fear in his prey, but Putnam responded by patting his mouth with his hand and yawning. Valdez reacted with beet red humiliation.

Putnam eventually took the offensive, antagonizing and teasing Valdez with his very own words.

"Oh, no. I hope it's not today."

"Can you give me a hint soon, Val? I must work it into my calendar."

"Will it be sometime this century?"

Putnam went on the offensive. He changed his sentence structure whenever talking to Valdez to include words like lame, feeble, flimsy, unfulfilled, impotent, failure, and disappointing. For the first time in his life, Putnam felt his size.

Waking from a quiet night of no emergency responses, Putnam sat on the edge of his bunk and stretched. It was a beautiful morning. He took a long, hot shower, whistling the whole time he walked back to his locker. Tousling his hair with a towel, rounding the corner of his locker row, he paused for a moment when something in Valdez's locker caught his eye. First, he double- checked each row of lockers to assure the room was empty. He returned to the Me-Shrine and, as usual, Valdez's locker door was open blocking the aisle.

Putnam stood in front of the Me-Shrine rubbing his towel on his wet hair, scanning across the usual collection of photos and news clippings and the *Great Meal* autograph from the Governor. Then, he noticed a recent addition– a bow; the red and white-striped bow from the top of the cookies that Evelynn had brought. *What does it mean?*

He closed the Me-Shrine door to step to his locker and reopened it once he passed. Standing behind the open Me-Shrine door gave him the seclusion of the entire locker aisle. The front of the locker door facing him was plain, not offensive in the least, other than the nametag *VALDEZ*.

Whistling halfheartedly a tune he made up himself, Putnam finished putting on his civilian clothes. While he buttoned his shirt, he strolled to the window at the end of the locker corridor. The sun was shining. The leaves on the trees were shivering in a light breeze. It was going to be a beautiful day. He returned his thoughts to the striped bow. He figured it was just another facet of Valdez's hollow threats – *just another psych-job and nothing more.*

Putnam slapped his cheeks with aftershave and grabbed his shoes and socks. He'd put them on downstairs. He locked his locker and replaced the key on the hook beside the door. He grabbed the brass handle to Valdez's locker door, closed it to leave and came face-to-face with Richard Valdez.

Blocking the entrance to the locker row, Valdez stood smirking. He held a galvanized bucket. He sloshed the pale around and tipped it slightly so Putnam could see the ice cubes floating in the swirling, purple-dyed water.

"Today's the day," sneered Valdez. He stepped back pulling the pail to his side, then reversed the motion toward Putnam who stood trapped in the small corridor. Instinctively, almost unconsciously, at the same instant Valdez threw the water, Putnam stepped back and pulled on the Me-Shrine door handle, shielding himself with Valdez's locker door. Hidden safe and dry behind the Me-Shrine door, Putnam missed the following sequence of events:

The sneer on Valdez's face quickly changed to an open-jawed, wide-eyed expression of terror. His arms stiffened and he arched back as he tried to reverse the forward momentum of the pail. He was successful in halting the pail, but not the purple water.

The wave of purple water slapped loud against his locker door, deluging the Me-Shrine. Ice cubes clunked hard against the door and bounced to the floor. Some photos tore loose and were washed away in the torrent of purple water. The remaining pictures were maculated with drips that raised their finishes and distorted their images with purple spots. The small picture- button that pinned the Girl Scout calendar to the door channeled the cascading water into an inverted V, deflecting it down each edge of the calendar and distorting the edges of the Governor's inscription.

From Putnam's perspective, it was not so much the sound of the splashing water that made him curious, but the sound of Valdez's scream, which was the loudest scream Putnam had ever heard anyone make with an inhale.

The metal bucket dropped from Valdez's hand as he rushed to the Me- Shrine. He dabbed at the photos and other memorabilia with his drenched and stained shower towel. He inadvertently trampled the photos and the striped bow bobbing in the puddle at his feet. Unconsciously, Valdez released another inhalation scream. Frantically, he dabbed and blotted his towel at the writing on the Girl Scout calendar. His towel turned purplish black with dye and the ink from the Governor's soluble inscription.

When the locker door finished dripping, Valdez stood back to survey the carnage. The photographs that still hung were curled and splotched purple.

The purple garter was ruined, too. Valdez scanned the Girl Scout calendar, the centerpiece of the Me-Shrine, hoping to find the Governor's memento unscathed. Most of the Governor's writing was obliterated, only the center of the salutation remained legible—*EAT ME*.

CHAPTER 15

THE PIONEER HOUSE

Brookfield—the name was once appropriate for a small dairy town in the rural lands outside Los Angeles, but the malignant growth of the surrounding metropolis methodically permeated Brookfield's pristine countryside. The crystal clear, spring-fed brook that had meandered for eons through the rolling hills of Brookfield was plotted and grid and entombed in cement as a permanent tributary to the L.A. storm drain system. The green dairy pastures were slowly consumed by the encroaching development and disappeared forever when farmer Willaset sold his dilapidated dairy to provide room for the Brookfield Industrial Park.

The only remnant of old man Willaset's dairy farm, and Brookfield's legacy, was the hay barn. It was a sturdy, log structure that the City of Brookfield claimed as a historical landmark and renamed it, *The Pioneer House*. Although no pioneers ever lived in the barn, except Mr. Willaset himself when the missus was angry, the citizens of Brookfield could not shake the romance of the name and planned to refurbish the barn into a community center. Renovation of the Willaset hay barn into an *all-purpose meeting facility* required substantial funding. Unfortunately, a stagnant economy and high vacancies at the Brookfield Industrial Park made the renovation a burden. During the final planning, many of the amenities were streamlined. The fabulous gourmet, restaurant-style kitchen was pared down to a two-burner stove and a barbecue on the back patio. The full scale, stadium-sized bathrooms were downsized to one toilet for both sexes to share. The state-of-the-art fire system, complete with smoke detectors, heat sensors, sprinkler heads and around-the-clock monitoring, was cut back to one manual pull box on the outside wall.

The Pioneer House was no longer nestled in vast fields of green grass, but instead sat awkwardly amongst the cold cement tilt-up warehouses of the Brookfield Industrial Park. At first, the Girl Scouts tried to conduct meetings there, but the scarcity of bathrooms for 50 screaming girls was an insurmountable obstacle. The Brookfield Senior Pioneers played bingo in the Pioneer House every Wednesday morning. With most members wearing bladder control briefs, the single bathroom was not an inconvenience. But, for the most part, the Pioneer House lay in state, forgotten in the shadows of the industrial buildings and the old Lutheran church directly across the street. The partially occupied industrial units were moderately busy during normal business hours, but after five o'clock the Pioneer House sat amongst darkened company. It was the perfect environment for a mischievous child like Glenn Turner to create the *Third Alarm Dash.*

For Glenn Turner, the Third Alarm Dash was more than a childish prank; it was fundamental to his life. The *Dash* proved his courage and hid his insecurities. The Third Alarm Dash symbolized his maturity—the metamorphic change from awkward adolescence into awkward puberty.

Besides being the mayor's grandson, Glenn Turner stood out from other kids in more uncomfortable ways. He had a thin, skeletal stature, liver-colored freckles splotched his pasty white complexion, and his head was anointed with wavy, parrot-red hair that destroyed every comb assigned to it. He would have stood out at a circus.

His insecurity was more than self-inflicted. At school, his peers teased him relentlessly. He was ridiculed because his basketball skills were not on par with his exaggerated height and he was teased for his neon-red hair. The 14-year-old's self-confidence weighed heavy, long after the last school bell. His release and only escape from the pressures of school was the Third Alarm Dash.

Valdez and Putnam hadn't spoken since the Me-Shrine incident. Everyone felt the chill. The station mood was further stressed by a plague of idiotic emergency calls. First, an old lady reported a structure fire because she saw steam coming from her neighbor's dryer vent. A man called 9-1-1 because he needed a battery changed in his smoke detector. After that, the Brookfield Convalescent Hospital called 9-1-1 because their mascot cat got its head stuck in a soup can; afterward, all the diners in the dining room were having chest pain.

As exasperating as the incidents were, it was their last call that irritated the crew most. Apparently, a lady dropped some change into the toilet, then dove after the sunken treasure, forcing her fist around the drain elbow. Like a large Chinese finger trap, the toilet seemed to tighten around her arm. With no way to grease her swollen, numb extremity, the firemen were forced to break the porcelain bowl to free her. Angered by the damage, she shook her fist of wet coins at the firemen and cussed them mercilessly and followed them out to the street yelling the whole way.

"There must be something in the water," said Valdez, as he slammed the door to the paramedic truck. "Or is today, *Idiot Day*?"

Razmunson nodded in agreement. "We should have used a toilet plunger and just flushed the old bat."

The excessive run load preempted Putnam's cooking routine. He had ordered pizzas delivered for lunch, but because of the extended time freeing the lady from the crapper, the firemen were going to eat cold pizza again for dinner, worsening their dour mood.

With no friends to hang around with, Glenn Turner busied himself watching his father smoke cigars. He studied his father roll each cigar between his fingers and blow out clouds of smoke, occasionally making smoke rings. For years, Glenn emulated his dad by going through the motions, feigning smoking his pencil or tootsie roll when no one was looking. Smoking a real cigar was important to Glenn because it allowed him the opportunity to revel like an adult after performing one of his juvenile pranks.

Earlier in the day, Glenn's father was called to an urgent sales meeting out of town. He didn't have time to finish the Cuban cigar he smuggled in from Canada during his last business trip. Glenn spotted the mid-length cigar in the ashtray on the patio table. He snatched it before his mom cleaned up and threw it away.

Holding up the stogie against the afternoon sun, Glenn Turner admired his latest acquisition; a real Cuban cigar, actually about half of one. Once he trimmed off the chewed end, it was about one third of its original size, but it was one hundred percent Cuban. And, as an additional stroke of luck, he found a nearly new disposable lighter lying in the street. Though the lighter was pink, it was in near perfect condition, probably run over only once or twice. He spun the lighter's flint wheel with his thumb. The chrome head was loose, but it lit and the flame was still adjustable. He shook it by his ear listening for the butane level. *Full*. Glenn Turner could not believe his good fortune—a Cuban cigar and a new lighter. It was definitely his lucky day.

After dusk, Glenn Turner and his retinue of misfit friends met in the dark parking lot of the Lutheran church across from the Pioneer House. Seven boys banded together with a common disposition. They all suffered from the cruel tricks of Mother Nature. Some of the boys were short, some fat, some with big noses, some with faces maculated with acne and most had several of the aforementioned ailments. The king and leader of the pack of adolescent oddballs was Glenn Turner.

The boys tightened their circle around Glenn as he held up his two prize possessions. They "oohed" and "ahhed", then carefully passed each article around their circle. Each boy smelled the cigar, like they watched Glenn do, with no idea what they smelled for, and nodded their approval. Then each one held the pink lighter, joked about the color, fumbled with the loose top piece, gave it a flick and adjusted the flame. When the items had traveled full circle, Glenn Turner slid the lighter into his shirt pocket and gave the cigar another stellar whiff before placing it alongside the lighter.

The custom before smoking the cigar was to perform the Third Alarm Dash. The rules were simple. The boy who felt the bravest snuck over to the Pioneer House and pulled the fire alarm. After activating the alarm, they hid in the junipers of the Lutheran church and waited for the fire Department to arrive. After a series of

Third Alarm Dashes, they began to recognize the firemen. Hiding so close, the boys could read the names stenciled on the turnout coats, but preferred their special nicknames: Blondie, Baldy, Spike, Lurch, and the Dwarf.

Dispersing from the church parking lot, the boys crawled through the junipers until they came to the outer edge facing the Pioneer House. They crouched motionless, while Glenn Turner stuck his pasty face out of the hedge and studied the surrounding buildings. The looping street was obviously deserted, but Glenn emphasized his surveillance technique, squinting his eyes and locking his stare on phantom sounds. The other members of the Third Alarm Club huddled in admiration.

Glenn raised his finger to his lips, "Shhhhhhh…" He walked hunched to the edge of the curb, looked right, then left, then ran across the street. He flattened against the log wall of the Pioneer House, the alarm box just across the porch. He leaped from the shadows and stood motionless under the red fire alarm sign. He made a grab for the alarm handle and stopped. He pulled back his hand and buffed his nails against his chest. He danced about capriciously, knowing it was increasing the anticipation of his audience in the hedge. He smashed the small glass cover and pulled the handle. The alarm bell clanged instantly. He sprinted toward the hedge. His lanky stride and large feet gave him speed, but also severe awkwardness. In one ungainly bound, he leaped from the porch, over the stairs and landed on the edge of the last step, sending him stumbling forward. His feet flapped desperately, trying to catch up with the rest of his body. Reaching the street, almost in control of his gait, he stubbed his toe against the high curb sending him skimming into the street on his chest. He bounced to a stop, quickly stood and dashed toward the hedge saying, "Shit!" a whole bunch of times along the way.

Diving into the darkened seclusion of the junipers, Glenn Turner patted his hands across his chest. He pulled out the pink lighter and the battered cigar from his torn shirt pocket. "Shit!" he repeated.

Under the drone of the clanging bell, the boys crouched in the hollows of the junipers, tittering and shooshing each other. In silence, Glenn Turner examined his bent cigar. Its outer wrapper was cracked and unraveling. Conceding to the damage, he hunkered down with the other boys and waited for the arrival of the fire engine.

"Re-heated pizza," said Valdez, disgusted as he poked at the pizza with his finger. The disappointment was mutual among the firemen, except Putnam who gnawed on a slice of cold pizza while he watched his other slices revolve in the microwave.

Ding...

Putnam quickly pulled the hot pizza from the microwave and fumbled with it, inverting one slice on top of the other, forming a pizza sandwich. He stuffed most of it into his mouth but couldn't close his lips when he chewed.

Ddddrrrriiiinnnngggg!

The dispatcher announced a manual pull alarm and a full structure assignment to the Pioneer House. Barker cursed aloud.

"Re-reheated pizza and now this," said Valdez, sliding his pizza into the trashcan. "This is turning out to be one hell of a day."

"This is your best meal ever, Terry. You must give me the recipe," added Razmunson.

Putnam poked the rest of his pizza into his mouth as he headed for the fire engine.

Because of the alarm history of the Pioneer House, Cole cancelled the other responding units. Engine 61 and Squad 61 would respond alone. Flipping on their red lights and throttling their motors, they filled the apparatus room with diesel exhaust and headed for the Pioneer House.

As usual, there was nothing unusual about the Pioneer House. The alarm was ringing but nobody was in sight. The paramedics checked the interior through the windows. Putnam reset the handle, silencing the deafening bell. Cole reported another false alarm to the dispatcher.

From the recesses of the hedge, the boys watched, silently studying the details of the firemen. Tonight, it was Blondie and his crew. They saw the Dwarf primp and fluff the helmet straps from his crushed pompadour. They watched with hilarity as Baldy flipped long strands of hair off his left shoulder and folded them back onto his shiny head. When the taillights of the fire engine disappeared down the street, the junipers erupted with cheers and kudos. It was thrilling to be so close to the firemen; so close to getting caught. The boys scrambled from the bushes into the dim parking lot of the church and gathered in a circle. In their center, Glenn Turner pulled the pink lighter and the frayed cigar from his pocket. "It's cigar thirty," he said, pretending not to notice the damaged goods. He raised the cigar above his head for all to see. No one spoke. Their attention was on the cigar. Glenn's attention was on them. Nobody noticed the crack in the lighter.

The gaggle of boys migrated to the back of the church and sat on the steps, watching Glenn as he wrapped his mouth around the stogie. The cigar's dark maduro wrapper was accented against his thin pale lips. He brought the lighter to the cigar tip. He took a couple of trial draws, but air drafted from the cracked sides, making the cigar difficult to light. His cheeks caved with each pull. He paused uncomfortably, feeling the expectation of his audience. Exasperated by the stubborn cigar, he set the lighter to *high*.

He spun the wobbly flint wheel, producing a tall flame and held the lighter steady as he tried a variety of lighting techniques. First, he rotated the cigar in the flame. Next, he pointed the cigar down into the flame. He could see the cigar tip starting to glow. *Just a little longer*, he thought. His thumb was numb from holding the lighter's valve. *Just a little longer*. The lighter's top was getting hot. *Just a little longer*. Warmed butane oozed from the crack. The chrome top fell off, the flame touched the butane and the lighter exploded. Reflexively, he released the lighter and gasped, but not before his face was engulfed in a fireball.

The flash was over in a second. Surprisingly, Glenn thought, the flame didn't feel that hot, just real warm, real fast. From where he stood, everything looked okay. His shirt was clean and intact. He raised his right hand to his face. His palm was chalky white, like it had been left in the bathtub too long.

"You okay, GT?" his friends asked as they gathered closer but stopped short of touching him. Glenn touched the sides of his face. His cheeks were moist and tender. He swept his fingers across the top of his head, pushing mounds of powdered hair down the back of his neck. Breathing anxiously, he tapped his fingertips on his leathery cheeks and felt sheets of skin draping from his chin. One of the boys ran for the alarm.

The apparatus room echoed from the engine's backup tone when the fire alarm sounded. Barker augmented it with profanity. Cole snatched the radio microphone from the dashboard clip and allowed his voice to convey his anger. "L.A. Engine Sixty-one. This location has a history of false alarms. Cancel other units. Engine Sixty-one will investigate—again." The dispatcher calmly repeated his orders and cancelled the incoming units.

After a choice selection of cuss words, Barker asked Cole, "Do you want to just drive over? Or go code? No sense in busting our butts running code for what we know is another false alarm."

Cole agreed. He told the paramedics to remain available in quarters. Barker shifted the fire engine into gear with a disgusted shove. Angrily, they turned down the street and headed to the other side of town, back to the Pioneer House.

Pulling through the weather-stained pillars guarding the entrance to the Brookfield Industrial Park, they heard the alarm clanging just like it did 20 minutes earlier. The narrow beams of the old fire engine's headlights softly illuminated the bland cement walls of the tilt-up warehouses that passed along the darkened street. Up ahead, around the sweeping curve, the Pioneer House came into view. But this time, in the shadow of the Lutheran church, they saw emerging figures, all waving their arms.

The fire engine pulled alongside the boys. All were featureless silhouettes against the backlight of the Pioneer House. Cole grabbed his flashlight and hopped down from the engine, eager to talk to the boys. "Are you the one's pulling the alarm?" he demanded, then stopped abruptly. In the swaying illumination of the flashlight beam, shadows from sagging blisters danced on Glenn Turner's face. The top of his head was smooth and sheen. His eyebrows were gone. His eyelashes were gone. His cheeks were pulled smooth and taut. His appearance took on an alien quality.

Cole fumbled with his radio mic. "L.A., Engine Sixty-one. We have a burn victim. Respond Squad Sixty-one, an air squad and a sheriff."

Barker ran to the back compartment of the fire engine for bottles of saline. Putnam connected an oxygen mask to the O2 bottle. The hiss of escaping oxygen muted as Putnam strapped the mask to the boy's face.

"What happened?" asked Cole.

"A cigarette lighter blew up in his face," blurted the other boy, standing behind Glenn Turner.

"Let him answer the question," admonished Cole, not taking his eyes off the burned boy. Turner described the explosion. His voice was gruff and raspy. He jumped when Barker sprinkled cool saline on his burns.

Putnam listened to Turner's chest with a stethoscope. He heard wheezes in the boy's upper airway. His breathing grew rapid and labored. He spoke in short sentences, saying his chest was getting tight. The taut skin between his clavicles and trachea drew in with each inhalation. Putnam took Cole's flashlight and closely examined the boy's face. His nasal hair was gone. Next, he visualized inside the boy's mouth. The roof of his mouth was red, and the back of his throat was swollen.

Barker cut holes in a wide gauze bandage and fashioned it into a mask. He moistened it with saline and draped it over the boy's face. They put a larger sheet over his neck and shoulders, then moistened it with saline. His exposed nerves stung as they reported the burn.

Sirens wailed in the distance as Glenn Turner wreathed in agony. The oxygen mask bit into his skin as it pressed against his raw nerves. Putnam loosened the straps and held the oxygen mask just off Turner's face. The boy's wheezing was louder than the gush of oxygen escaping from around the open mask.

Squad 61 screeched to a stop beside the fire engine. Officer Parker was behind them. The paramedics heard the boy's shrill wheezing as they trotted up. Captain Walker gave them a brief synopsis of the boy's condition, although the audible wheezes said it all. Valdez dialed up the hospital. He pressed the cellular phone between his ear and shoulder as he wrote on the clipboard.

"Our Lady of Brookfield, this is Squad Sixty-one. We've got a 14-year- old male suffering from second and third degree burns to the face, neck and respiratory tract. The patient is alert and oriented, but he has severe respiratory compromise. We want standing orders for dual lines of normal saline, morphine and continued oxygen therapy. Plus, an order for endotracheal intubation. Over." His head nodded as he wrote.

While Valdez spoke to the hospital's MICN, Razmunson rolled out the burn pack beside the boy. Putnam sprinkled the sterile sheets with saline. The three firemen lifted the boy onto the burn pack and wrapped his torso in the cool sheets. Putnam continued irrigating his face and neck with water, while Razmunson opened his airway kit. Glenn Turner's anoxia made him anxious. He struggled against lying down. Each breath increased in pitch, sounding more and more like squeaks. His trachea was swelling closed.

Still on the cell phone, Valdez waved two fingers at Razmunson, then tapped his forearm, signaling two IV's. Then motioned his finger to his throat, indicating hospital approval for endotracheal intubation.

Razmunson acknowledged with a quick nod. He spun to his airway kit and removed the laryngoscope and the 13-inch endotracheal tube. He snapped the curved blade into the laryngoscope handle. The optic light flashed on. He slid the thin flexible stylet into the hollow of the endotracheal tube and formed it to the curve of the boy's larynx. Kneeling, Razmunson straddled the boy's head. He looked at Captain Walker. "What's his name?"

"Glenn Turner...14," interjected Valdez, reading from the rescue report form.

Razmunson leaned over the boy, whose head was positioned between his knees.

"Glenn, your windpipe is burned and is swelling shut. I have to put a tube down your throat, while I still can. It's difficult and it's not fun. You have to work with me."

The young boy listened intently, struggling to hear over the shrill of his own breathing.

"When I put this tube in your throat, it'll make you gag, but fight the urge to vomit. We have to do this to keep you breathing. Okay? Ready?"

The whistle-like breathing paused for a moment as the young boy grasped the importance of the situation. *I could die.* Through his reddened skin, he looked pale for the first time. The paramedic's words confirmed the severity of his injury. His shoulders shook, while the whistling turned into trills as Turner started to cry.

Razmunson sprawled prone on the grass above the boy's head. He scooted back, first face-to-face, then forehead-to-forehead with the youth. He was positioned to look down Glenn Turner's trachea.

"Open your mouth when you're ready, Glenn."

The scared boy opened his trembling mouth. The light of the laryngoscope blade illuminated his red and bloated mouth cavity. The swollen entrance to his throat was reduced to a mere slit. Razmunson slid the tip of the laryngoscope blade into the tight passage of the boy's throat. Glenn wretched and gagged. Razmunson pulled back the laryngoscope. Glenn Turner contorted and arched as he twisted to vomit. Vomiting would fatally clog his airway.

"Fight it, Glenn! Don't hurl," ordered Razmunson, as he waited anxiously for the boy's nausea to subside. "You've got to fight it, Glenn. We're running out of time."

The young boy nodded, lay on his back again and opened his mouth.

Again, Razmunson slid the blunt blade through the narrow passage of the throat, again stimulating the boy's gag reflex. Glenn wretched and arched but held steady. Pushing the laryngoscope blade past the throat opening, into the larynx, Razmunson visualized the vocal cords. The white cords sat opposed to each other above the opening to the trachea. The trachea appeared undamaged. Razmunson slid the clear endotracheal tube down the channel of the laryngoscope blade, past the epiglottis, between the vocal cords and into the trachea. He inflated the bulb at the end of the tube, wedging it at the entrance to the lungs just above the corina.

"Breathe, Glenn," said Razmunson. He listened for air passing from the tube's end. Air rushed in and out until Turner relaxed. The blue pallor of his lips returned to pink.

Razmunson spun around on one knee to Glenn's side and looked him in the eye. "Breathing better?"

Unable to speak because the endotracheal tube separated his vocal cords, Glenn could only nod. Air rushed in and out of the tube as the young boy cried, sounding much like a bugling elk.

Secondary to the boy's compromised breathing, he was in jeopardy of hypovolemia and shock. "Now, I've got to start some IV's, Glenn," said Razmunson, encouraging the young boy with a squeeze of his arm. "But it won't be as bad as the tube, I promise."

While Razmunson established the IV's, the surrounding air was filled with the pounding of helicopter rotors. Officer Parker directed the Los Angeles Fire copter into the church parking lot. Its rotor wash blew a gale of napkins, Styrofoam cups, and donut bags from Parker's open patrol car. Two air-squad paramedics ran hunched from the helicopter with their stretcher. Razmunson finished taping the second IV just as they arrived. The four firemen lifted Glenn Turner onto the Miller board where he was strapped in and rushed to the waiting helicopter.

"They'll take it from here, Glenn," said Razmunson, yelling over the rotor's beat. "It'll be tough, but you hang in there." He gave the young boy's hand another squeeze.

Glenn Turner waved a pitiful goodbye to Razmunson as the airmen closed the copter door. Little pieces of asphalt shot at the firemen, stinging their cheeks and bare skin, but defiantly they watched the copter lift off and disappear into the distance.

There were no more false alarms at the Pioneer House.

CHAPTER 16

THE MAN IN PAJAMAS

Tuesday morning when Putnam arrived at the firehouse the offgoing crew was still rousing from their busy night. He rummaged through the refrigerator before going upstairs. Biting into a cold pancake rolled round a glob of peanut butter, he pushed through the kitchen door and noticed the paramedic truck on the front apron. Valdez was washing it. Putnam forewent wishing Valdez a good morning, not sure of how it would be received.

Putnam trotted upstairs, past the dorm. Some of the bunks still contained firemen sprawled under the sheets. As he passed the bathroom, he casually waved to the two firemen shaving at the sinks. Continuing to the locker room, he had a surprise. He was startled by the bright sunlight flooding through the window at the end of the locker aisle. Valdez's locker door was shut. The Me-Shrine was closed. Putnam doublechecked the locker room to confirm that he was alone, and then inquisitively opened the Me-Shrine.

The inside of the door was barren, except for pinholes and residual pieces of tape. There was a single shoebox on the bottom shelf of the locker. He lifted the box cover and found oddments of the Me-Shrine. Curled and stained photographs, torn and deteriorated newspaper clippings and a moldy purple garter belt. The bottom picture torn from the top of the Girl Scout calendar was folded to fit the confines of the shoebox. Putnam unfolded the calendar and read the remnants of the Governor's inscription: *EAT ME.*

He blurted out a laugh before covering his mouth with his hand. He checked down the hallway. The firemen were still talking in the bathroom. He read the calendar again and checked its sheen against the light for any hint of the Governor's original message. He saw nothing, just *EAT ME*. He returned the calendar to the shoebox and gently replaced the lid.

Putnam felt a twinge of pity for Valdez. The destruction of the Me-Shrine had to be devastating. He decided he wouldn't tease Val anymore. As he closed the locker door, something familiar caught his eye. On the second shelf sat a red and white striped bow still creased with the imprint of a boot sole. A shiver of fear ran the course of his neck. He wondered, *Did things go too far?*

The past few days had been nothing but a series of calamities for the soft-spoken accountant. The trouble began when he brought his male friend to the office party and announced their engagement. The following Monday, he lost his job under the guise of "belt tightening." He had to postpone their nuptials. Yesterday, his fiancé missed his third AA meeting. And again, they had argued most of the night.

Startled, he sat up alone, saturated in sweat. He slid from the bed into his slippers and ventured into the dimly lit den, where his partner, anger, and a near empty bottle of whiskey were waiting.

Later that morning, when Putnam approached the paramedics for dinner money, he was careful to keep his conversation strictly business with Valdez. "Who's in for chow?" asked Putnam, forcing a deliberate smile.

"Here's my money," said Razmunson, unfolding his wallet. "Will you guarantee there'll be flavor this time?"

"The only thing I guarantee is that it'll make a turd," returned Putnam.

"Before or after we eat it?" asked Valdez, holding out his money.

Putnam didn't know how to take Val's comment, which was sarcastic in a playful way—unlike Valdez. Putnam took the money with a grin and hopped on the engine. On the way to the store, Engine 61 was dispatched to a four- year-old boy found crying in his front yard. The previous day in preschool he had learned how to find his pulse. At home, while his mom showered, he tried to find his pulse but couldn't. So, he called 9-1-1 because he thought he was dead.

Captain Walker cancelled the incoming squad and ambulance. Putnam calmed the boy and showed him how to find the soft thumping pulse at the end of his wrist. Barker instructed the boy on other ways to determine if he was alive, like blowing condensation on a mirror or pinching your inner arm until your eyes tear.

Within 15 minutes, they were back in the store. Somewhere in the vegetable aisle, they were dispatched to an electrocution. When they raced into the house, they met the victim of the electrocution who explained that she had dropped a disposable camera on her foot causing the flash to discharge. She called 9-1-1 because she thought she had been electrocuted. Barker started to educate the woman on electricity and common sense, but Cole wedged between the two and explained the situation to the lady in less caustic terms.

Once again, Cole and his crew returned to the store. They located their shopping cart of spongy frozen vegetables and warm packages of meat. Shortly after, they were dispatched to an apartment house fire. When they arrived at the apartment, an angry woman met them at the door and demanded a pen.

"You want a pen?" asked Cole.

"My roommate is having a tizzy because I'm a few days late with my half of the rent. I've got the money, but I need a pen to write a check. She said, 'it's an emergency'. Since firemen handle emergencies, I called you guys."

"You needed a pen, so you reported a fire?" asked Cole.

"I knew you guys would come fast. And here you are. Can I use your pen? Please."

"Hold that thought," said Cole. He pulled out his portable radio and cancelled the responding four fire engines, two ladder trucks and paramedic squad, then politely loaned his pen to the woman. After she wrote her check, Cole loaned his pen to Officer Parker who wrote her a citation.

The engine crew did not return to the store. Instead, they ordered pizza delivered to the firehouse and hoped they'd be in quarters when it arrived. The crew waited patiently in the recliners for the pizza man. They discussed the day's runs. Barker could not believe how many morons lived in Brookfield. He concluded, "It's got to be the idiot capital of the universe."

The doorbell rang.

"I'll get it," said Valdez, springing to his feet.

"It's too soon for the delivery boy," said Putnam, just hanging up the phone from the pizza parlor.

"You stay put, Putz. I'll handle it," said Valdez as he pushed through the kitchen door. Putnam was perplexed. Such polite spontaneity was unusual for Valdez.

"Is he expecting a package?" asked Razmunson, also puzzled by Valdez's enthusiasm.

Putnam eased himself back into his recliner. For the first time, somebody else was answering the door. The TV program was ending about the same time Valdez backed through the kitchen door with a plate of cookies.

"Where'd you get those?" asked Putnam, alarmed.

"Oh," said Valdez, nonchalantly. "The Cookie Girl."

"Why didn't she come back to visit us?" asked Putnam, now standing.

Deliberately ignoring the question, Valdez dug into Putnam with his observation, "The Cookie Girl is looking pretty good lately. Did she do something with her hair?"

"Why didn't she come back to visit *us*?" demanded Putnam.

"Well," began Valdez, as he slowly disrobed the cookies of their cellophane cover, "she was running out of time, I guess. We didn't discuss that." He knew exactly what he was saying as he rummaged through the refrigerator. "We talked so long, we lost track of time." He turned from the fridge with a carton of milk and set it on the table, then pulled the colorful bow off the cellophane and stuck it to his uniform just above his nametag.

"The Cookie Girl is going to college next semester," he said, pensively dunking his cookie into his glass of milk. "She may be getting her own apartment. I said I'd help her look for one." Putnam stood silent and obviously tense, loathing Valdez's dramatic embellishments.

"You know," said Valdez, speaking over the rim of his milk glass. "I'd like to see more of that girl."

Putnam's skin flushed red. He did not partake in the cookies with the rest of the crew. Instead, he stormed up to the dormitory and sat on the weight bench. He feverishly pumped the barbell. The weight that Valdez had trained the last three months to press Putnam pumped with effortless repetition. His workout was interrupted when Valdez strolled into the dorm, wearing his faded sweatpants and torn tee shirt that accented his well-defined muscles.

"Nothing like a workout to work off those chocolate chip cookies, right, Putz?"

Putnam did not answer. He stood from the weight bench and stared out the window at the stark wall of the warehouse next door.

Valdez took his place on the weight bench and continued his assault. "The Cookie Girl came at an opportune time, since I'm on the skids with my main squeeze. I'm looking for a temp." His comment spun Putnam around like a pull on the shoulder. Valdez pretended not to notice Putnam's anger as he continued. "She's not really my type, but I figure *any port in a storm*, right, Putz?"

The last time Terrence Putnam lost his temper was in the sixth grade. It scared him then. It was scaring him now. But when Valdez stood to change the weight discs on the barbell, Putnam stepped directly in front of him. "Leave Evelynn out of this," he said, inches from Valdez's face.

"What do you mean?" asked Valdez, innocently touching his hand to his own chest. His left cheek rippled as his skinny little mustache tweaked into a sinister smirk.

Losing his last bit of restraint, Putnam gave in to his anger and shoved Valdez against the wall. He pressed his right forearm against the little man's neck, grabbed his flimsy tee shirt and slid him up the wall. Valdez had to use both hands and all his strength to press the big man's arm off his throat.

"Let go, Putz," Valdez demanded.

Putnam's red face and crazed expression were demonic. His eyes were glazed. Tears ran down his cheeks. He said nothing, pressing harder against Valdez's throat. He could feel Valdez's Adam's apple move with each swallow.

"Let go, Putz," demanded Valdez, his voice raspy, as he desperately pressed against Putnam's thick forearm. He stretched his toes down but couldn't touch the floor. He stared helplessly into the eyes of the monster he'd created. Valdez was in a position to kick Putnam in the groin but did not want to risk the retaliation. He hung in silence.

Finally, Putnam spoke, "If you have a problem with me, handle it with *me*. Leave the girl out of it." He leaned his massive head forward, leaving only inches between their two faces. His breath was hot and smelled of onion. He bumped his forearm into Valdez throat.

"Let go, Putz," Valdez retorted. His face was purple, and each breath began with a wheeze.

Ding…

The big fireman held the little fireman without flinching at the alarm. He glared at Valdez, who motioned toward the bell with his eyes. Slowly, Putnam slid him down the wall. Valdez felt relief when his toes finally touched the floor.

Ddddrrrriiiinnnngggg!

"Engine Sixty-one. Vehicle fire. Corner of Maple and Elm. Thirteen Twenty-nine."

"That's for you, Putz," said Valdez, tipping his head toward the pole.

Putnam pulled Valdez forward by the shirt, and then shoved him back against the wall, releasing his grip. He turned and stepped to the pole hole. He leaned against the pole, hesitated, and then pushed himself to his feet. He pointed at Valdez, "My name is not

Putz. Understand, little man?" Not waiting for a reply, he wrapped his arm around the pole and slid down.

Once out of sight, Valdez released his pent breath and rubbed his tender throat.

Putnam jumped on the fire engine; Walker and Barker were already in their seats.

"It's on the other side of town," said Barker. "We'll be just in time to cool the chassis."

The bay door clattered as it rolled open, revealing a distant column of black smoke.

The man in paisley pajamas found it difficult to reason with his drunken lover. Their argument was escalating. One threatened intervention; the other threatened violence.

"I'm calling the police," said the man in the pajamas, picking up the phone. He punched *9-1-1* on the keypad, but before he spoke to the emergency operator the phone was jerked from his hand and beat against his head. The dispatcher sensed urgency from the succession of thuds before the phone line crackled and went dead. *DISCONNECT* flashed on her monitor as she dispatched sheriff and rescue units to the calling location.

Near the end of the block, Captain Walker saw the burning station wagon. Fire ravaged the engine compartment. Flames fingered from the cracks around the hood and grill and burned in fiery puddles under the car. The owner gallantly swatted at the flames with his sport coat. When he saw the approaching fire engine, he waved his smoking blazer. Barker passed the burning car to position the fire engine uphill of the burning streams of gasoline and up wind from the black, carcinogenic smoke.

Captain Walker pulled the attack line toward the fire, while Putnam donned his breathing apparatus and synched the straps to his facemask. At the engineer's control panel on the opposite side of the fire engine, Barker shifted the transmission out of *road* and into *pump*. The fire hose hardened with water pressure as Barker throttled up the motor.

Putnam ran to Cole, the inside of his mask glistened with condensation. Cole gave Putnam a quick inspection. He tightened his helmet strap and straightened his Nomex fire hood, leaving no bare skin exposed. Putnam took the nozzle and advanced on the fire. Putnam felt cool air rush across the nape of his neck as it was sucked into the venturi of the water spray. The dense wall of smoke gave way to the water, exposing the blackened fenders and dirty-orange flames. The vehicle groaned and crept like a wounded animal as shorting wires engaged the starter motor. The horn blew; all three firemen jumped.

Wearing his breathing apparatus and armed with a pry bar, Captain Walker ducked into the billowing smoke and forced the hood open. Starved for air, the fire leaped from the engine compartment, biting through his gloves. He dropped the hood. He pried up the corner of the hood again; this time, the fire retreated from Putnam's water spray. A rear tire exploded, dropping the car with a violent crash. Gasoline floated atop the running water and spread fire the length of the vehicle. Putnam pulled the fire hose to the rear of the car. Flames, fed by the shorting electric fuel pump, burned unchecked under the gasoline tank. The fuel tank bulged from pressurized vapor. A jet of burning gasoline vapor whistled out the gas cap vent hole.

Putnam knelt by the left rear fender and sprayed at the flames. He heard the gasoline pop and crackle as it boiled in the fuel tank. Flames fingered around the gas tank and burned into the fill tube's rubber connection. The vapor pressure of the boiling gasoline exceeded the integrity of the burning rubber and exploded into a giant fireball.

The explosion blew the vehicle two feet off the ground. Gasoline shot from under the car, enveloping Putnam in flames. He spun in terror; everything around him boiled red and orange. Lost for direction and stumbling backwards, he tripped over the fire hose and fell out of the fireball. His view cleared as the ball of flames rose into the blue sky like a nuclear mushroom cloud.

Putnam rolled back from the car and scrambled to his feet. He quickly took stock of himself. His turnouts were singed and smoldering, but not on fire. Each heaving breath filled his lungs with clean, cool air from his bottle. His facemask withstood the blast. He was alive.

"Are you alright?" shouted Cole, shaking Putnam.

Putnam nodded, dizzy. He swayed like a punch-drunk fighter, shook the haze from his head, picked up the hose and headed back to attack the fire.

At the firehouse, Paramedics Razmunson and Valdez groaned in the recliners after eating most of the two large pizzas left in their custody.

Ding…

"Sure," said Razmunson disgruntled. "Why couldn't the bell ring *before* we ate the pizza?"

Ddddrrrriiiinnnngggg!

"Engine Thirty. Squad Sixty-one. In Sixty-one's area. Assault. Seven-Three-Seven-Two Chestnut Drive. Cross Oak. Sheriffs responding. Thirteen- fifty."

Both medics moaned as they pulled themselves up from their supine position. Chestnut Drive was just a few blocks away. They'd be there in no time.

The response down Chestnut Drive was pleasant. The solid canopy of elm branches made Chestnut Drive shaded and cool even on the hottest summer days. Unfortunately, in addition to the elms, Chestnut Drive was known for its cavernous dips. Valdez sped down the block then sharply braked before each intersection. He gingerly eased across each dip, but still scraped the rear bumper.

Squad 61 arrived first on scene, at least five minutes ahead of Engine 30. On the front lawn of the modest little house sat a middle-aged man in bloodied pajamas. His hands were clamped tight to his head and his head was vised tight between his knees. Bright red blood oozed between his fingers. He did not look up when Valdez accidentally stepped on the siren button when getting out of the paramedic truck.

"He failed his hearing test," joked Razmunson.

Razmunson knelt next to the bloody man. The outline of his skinny frame showed through the moist pajama fabric that clung to his skin. "Sir," began Razmunson. "What happened?"

The man did not respond. He stayed motionless, his hands squeezing his bloody head.

Valdez scribbled hard on the rescue report. "What is your name, sir?" he demanded. The man did not answer. Valdez repeated the question, not hiding his irritation. There was no response.

Razmunson pulled the man's fingers from his bloody scalp. He examined the large avulsion, noting yellow clumps of fat and the creamy brilliance of the man's exposed skull. Most of the blood was coagulated and stuck to his latex gloves in jelly-like globs. Razmunson pulled on the flap of scalp, erupting it in profuse bleeding. He pressed a bandage compress over the flap and held it with one hand, but the adhesive tape refused to stick to the moist blood. He pressed hard on the bandage as he rifled through the drug box for a roll of gauze to wrap the dressing to the man's head.

"Sir, you need to help us, so we can help you," Razmunson said, using a more compassionate tone than his partner. He rummaged through the drug box for his blood pressure cuff, then turned his attention back to the man in the pajamas. "Sir, what's your name? Tell us how you got hurt."

The man raised his head drowsily. His left eyelid was swollen purple and closed over his eye. Razmunson pressed on the man's swollen cheeks, causing him to wince from the crepitation. The man's nose was bleeding and angled to the left.

There was a distant crashing sound that caused both paramedics and the man in the pajamas to look up. Racing down the street, traveling at a speed too fast for Chestnut Drive came Officer Parker, his siren wailing. The deep, throaty draw of the four-barrel carburetor betrayed his speed. In the darkened shadows of Chestnut

Drive, glowing sparks and pieces of hot metal bounced behind the speeding patrol car as it careened across each dip.

"His throttle must be stuck," said Valdez, transfixed in disbelief at the sparking patrol car.

"He probably got called away from a fresh batch of maple bars," added Razmunson.

Next, their attention was averted to the house by the squeaking hinges, and the piercing squeal of the screen door as it scratched against the barrel of a gun. Seeing the rifle barrel before his partner did, Valdez dropped the clipboard and ran in the opposite direction of the gun. Razmunson was squatting behind the man in the pajamas when he saw the rifle barrel emerging from the doorway. He stood and hoisted his patient by the armpits, but the bloodied man collapsed and fell back on top of him.

Running full speed in blinding retreat, Valdez slammed directly into the utility box of the paramedic truck. Frantically, he looked right, then left, then right again. Both directions looked miles long. He turned left, grappled along the side of the utility box and scrambled for shelter behind the truck.

Tires screeched as the sheriff car skid to a stop in front of the house. Officer Parker fumbled forever at the dashboard before emerging from the patrol car, shotgun in hand. The gun announced a chilling *clack* as Officer Parker pumped a round into the chamber.

Undaunted by Officer Parker, the man on the porch stood defiantly as the screen door slapped closed behind him. He surveyed the front yard, scarcely noticing the patrol car and the paramedic truck. Showing no expression, he turned the barrel of his rifle to the man in pajamas.

"Drop your weapon!" ordered Officer Parker to the man on the porch. His voice was unusually authoritative. The man on the porch did not respond to Parker's orders but held his bloodshot eyes on the bloodied man in pajamas and pulled the gunstock to his shoulder.

Officer Parker still held his shotgun low by his hip. "Drop your weapon or I'll shoot," he shouted again, while groping for the shotgun trigger. His fingers felt thick and clumsy.

The man on the porch put the man in pajamas in his sights. Unprepared, Officer Parker fired the shotgun from his hip. The gun recoiled sharply, bending his index finger backwards against the trigger guard until it snapped.

Buckshot splattered chucks of stucco and shattered the window behind the man on the porch. Several pellets struck his right shoulder, splattering his dingy white tee shirt with blood. He staggered back, struggling to keep his balance, then raised his gun again and reacquired his bead on the man in pajamas.

Officer Parker struggled with his injured hand and his discharged weapon. His broken finger made his entire arm throb. He squeezed the gunstock under his injured arm and pumped the handle back and forward left- handed, expelling the shell and reloading the chamber. He had to physically look at his right hand to find another viable finger. He thrust his center finger in the trigger, jamming his broken finger into the side of the trigger guard. The pain made him gasp aloud. He squeezed the gunstock against his hip and fired. The shotgun violently surged backwards. His center finger, trapped beneath the trigger guard, bent backwards against its joint and popped. He dropped the shotgun.

The main pattern of pellets struck the man in the right flank of his abdomen, spinning him full circle. Stray pellets struck peripherally around the man, dusting him with plaster. His gun fired into the air. He stumbled backward and slid down the stucco wall that rolled up his shirt, exposing his large white belly and protruding navel. He sat motionless. His left hand lay flaccid on his lap. The man, the porch, and the wall behind were flowing with blood. Slowly, the man on the porch opened his eyes and searched for his rifle. His right hand found it.

Officer Parker called for the man to drop his weapon. His voice lost its authoritative inflection, sounding more like a plea. The man on the porch never took his eyes off the man in pajamas, as

Razmunson dragged him across the lawn. With considerable effort, the man on the porch pulled the rifle across his thighs.

Officer Parker picked up the shotgun with his left hand. Chambering the next shotgun shell was agonizingly slow. He placed the gunstock between his knees, pumped the handle back then forward. Officer Parker stood up, holding the shotgun left-handed. He rested the gun's barrel across his right forearm; his right hand hung limp, fingers dangled in several oblique directions. He tightened his left index finger against the trigger. He tipped his head in line with the end sight of the shotgun barrel and called to the man on the porch to desist.

Struggling for breath, the man on the porch spun the gun on his thighs until it pointed at his lover.

The report of Officer Parker's shotgun echoed through the street. Pellets in the lower pattern tore into the man's legs shredding his pants. Some pellets struck his flabby abdomen causing it to ripple and quake. The main concentration of shot ripped into the center of the man's chest, lofting a red mist. The man on the porch slammed hard against the wall, then collapsed to his left side, smearing a bloody arc across the stucco.

Officer Parker lowered the shotgun to his side, then dropped it altogether. Razmunson kept scooting, dragging the bloody man in pajamas further away. The man on the porch lay still, gun on his lap, his left foot twitching in the pool of blood enlarging around him.

"Murderer!" yelled the man in pajamas, suddenly coming to life. "You killed him!" he screamed at Officer Parker. "Murderer!" He shook himself free from Razmunson and stumbled to the man on the porch. "Murderer! Murderer!" he repeated as he wept over his lover's body.

When the paramedics returned to the station, Captain Walker was waiting. He probed the two medics for signs of stress or emotional unraveling, but both paramedics claimed they were unfettered. Both declined Cole's offer for psychological counseling.

Cole suspiciously released them from his office. Razmunson went to the kitchen for a cup of coffee. Valdez went upstairs to shower.

The sound of running water prompted Putnam to go upstairs. It was after five o'clock, so he decided to change from his smoky uniform into comfortable sweats. Passing the bathroom on his way to the locker room, he noticed Valdez absorbed in deep thought. Wearing only a towel, Valdez stood hunched over the bathroom sink. His hair tousled, head and shoulders slumped, his weight resting on his forearms. A steam cloud rolled over the shower door. He was unaware Putnam was in the doorway.

"You okay, Val?"

Putnam's voice obviously startled Valdez, who straightened his posture and stared into the mirror. "Yeah, I'm alright." Knowing his voice was not convincing, he ducked into the shower.

Putnam stepped farther into the bathroom and took a quick glance under the toilet stalls to ensure that he and Valdez were alone. "Val, I heard about your run today. The man with the gun..." He inched forward cautiously, wondering what he would say next. He stood beside the shower door. "I know if I had been there, I would have been scared...real scared."

There was no response from the other side of the shower door. Through the frosted glass, Putnam could see Valdez motionless under the shower- head, letting the hot water drill into his shoulders.

"If...if you ever feel like talking about it, Val, I'd like to help..." Putnam took a few small steps back toward the door. "Any time," he said, not knowing what else to say.

There was no reply, just the sound of spraying water. Putnam took a few more steps backwards before turning to leave. The shower door cracked open. Valdez stuck his wet face out of the shower. "I'll keep that in mind."

There was an awkward silence. Putnam just nodded and shuffled backwards toward the door.

Valdez called him back. "Terry," he said, giving his next words deliberate thought. "Thanks. And that stuff I said about the Cookie Girl, I didn't mean it."

Putnam flashed a grin and left, leaving Valdez to shower alone with his thoughts.

On a barren stretch of the I-5 Freeway, three hours out of Oakland, Dennis Upman drove his diesel tractor south. The night was thick and moonless, only a dull section of highway was illuminated in his headlights. The large steering wheel felt heavy and pulled left. He decided to check the tires. He needed a break anyway. The meth he took earlier was wearing off. Downshifting and signaling right, he moved the big rig onto the gravel shoulder that rattled his truck to a stop.

He opened the cab door and sat on the seat. He could smell the blown tire. The bitter rubber smelled like hot tar. A flap of retread was wedged and smoldering under the fender. The overheated rim crackled and popped as it cooled in the evening air. He jumped from the cab and slammed the door. The brown plastic vial bounced at his feet. He stuffed the container into his jacket pocket.

The sky was pure black as if even the stars were hiding. The flickering flame of his cigarette lighter did little to illuminate under the trailer. "Well, I'm glad I've done this a hundred goddamn times, so I can do it in the goddamn dark!" he shouted into the darkness as he kicked at the gravel. "Shit!"

Before beginning work, he wanted to *get right*. He walked to the other side of the truck and pulled the vial from his pocket. His highs weren't lasting as long as they did at the beginning of his trip. He pulled a bottle of water from the truck cab and shook three pills into his hand. He swallowed them with the water. He lit a cigarette. The flame cast ominous, dancing shadows across his face. He casually blew smoke into the dark night and waited for the pills to take effect. "Much better," he said to the darkness. Wired with energy, he began his work.

Not wanting to scuff his leather jacket, he carefully draped it across the spare tire rack, while he wrestled with the blown tire. The frayed strands of hot rubber seared his bare hands as he pulled on the shredded tire. His jacket fell to the ground as he shook the spare tire from its mount. "God-damnit!" he yelled at the night, as he watched his coat crumple to the dirt. His hands were too greasy to pick up the jacket, so it would have to lay until he was finished.

The spare tire was here. Even in the cool night air, he perspired. Everything was a struggle. Aligning the lugs with the rim was a bitch. Groping for the nuts in the dark was a bitch. And torquing the nuts with a bent lug wrench pulled at something in his back. And to top everything else, he was losing precious time.

When Upman was done replacing the tire and his tools, he delicately picked his leather coat off the ground, carefully pinching the sleeve with the tips of his dirty fingers. He returned to the cab, not noticing the brown vial on the ground in front of the inside dual. The physical activity and his scraped and stinging knuckles angered him wide awake. He roared the diesel engine back to life and rolled the big rig forward, crushing the amphetamines and his plans to stay awake, as he headed south to the Torrance refinery.

CHAPTER 17

THE WALTZ

Cole sat on a picnic blanket in the Brookfield City Park, enjoying his day off. The lush greenery and golden evening sky would be romantic if he wasn't sitting with Putnam who tugged nervously at his tuxedo cuffs.

"Relax, Terry," said Cole. "The plan is going to work just fine."

"I'm sure *your* plan will work great," said Putnam. "I'm just worried about *my* part. I hope I don't screw it up."

"You'll do fine, Terry. Just be yourself. Okay?" Cole gave Putnam an encouraging wink and two thumbs up. "Now, let's get the show on the road," said Cole as he looked at his watch. He flipped open his cell phone and punched in numbers. He dialed up Dorothy Greene. "Dorothy. Cole. We're ready. You remember what to do? Good." He folded the phone and placed it in his backpack. He sipped from his iced tea and glanced again at his watch. "You better get into position, Terry. And don't worry."

"I hope I don't screw it up," said Putnam, pulling at the bow tie squeezing his thick neck.

"Just be yourself and leave the rest to us," said Cole, patting Pluggy, the remote-control fire hydrant. "Nothing will go wrong."

Putnam left and Cole strapped on the audio/video headset. He took another sip of his tea and gave the joystick a push, sending Pluggy scooting across the park grass to Dorothy Greene's house.

Cole deftly wove the little hydrant down the sidewalk to the crosswalk ramp, where he waited for a green light. He patiently sipped from his cup. Suddenly, with a hiss and a slap, Cole was hit with water. He yanked the headset off and found himself in the midst of the park's lawn sprinklers. He jumped to his feet, spilling iced tea on the remotecontrol unit. The control unit sparked. Pluggy lunged into the street.

Cole ran through the sprinklers and into the intersection, just as Evelynn drove past. Preoccupied, she didn't see Cole snatch the fire hydrant from the street and drop the control unit. Pieces flew everywhere. Evelynn raced down the street and swerved to the curb in front of her great-aunt's house. Cole gathered the scattered parts from the control unit and picked up the unresponsive hydrant. He hid behind a parked car until Evelynn disappeared inside her aunt's house. He set the plastic hydrant next to her car and ran for cover.

Evelynn rushed from room to room in her aunt's house and was surprised to see Aunt Dorothy merrily making tea in the kitchen. "What's wrong?" asked Evelynn, slightly out of breath.

"Oh, hello, deary. You must have had an exhausting trip. Have some tea?"

"I came over as fast as I could. You said it was urgent. So, I came quickly. Because you said it was 'urgent."

"What's urgent?"

"*You* said something was urgent," Evelynn insisted, waving her hands as she spoke.

"Sit down, child. You are too young to be in such a fuss." Aunt Dorothy handed Evelynn a cup of tea. "Sit, child…sit."

Cole ran down the street with the broken remote control. He ducked into a small, crowded café. "Where's your bathroom?" he demanded with his armload of hardware.

The waitress gestured with her tray of dinner plates.

Cole shoved open the bathroom door and dumped the pieces of the control module on the sink counter. He assembled the various pieces of the control module and flipped the power switch on and off, but the unit was dead. An old man meticulously dried his hands under the hand dryer. Cole waited impatiently, clearing his throat and tapping his foot. When the perturbed man left, Cole rotated the control unit under the hot air. He checked the box for damage and found that one battery was missing. He dumped the contents of his backpack on the counter and grabbed the AC power adapter. The bathroom had no electrical outlet. He ran into the restaurant. The waitress was walking by with a stack of dirty plates.

"Do you have an electrical outlet I can use?"

"We don't let anybody use electricity except customers," snipped the waitress.

"Look, here's a twenty," said Cole, fumbling with his wallet. "I'll buy something. Where's the outlet?"

"There's a thirty minute wait for a table," she said, snapping her gum and nodding toward the dining room.

"Please, this is urgent. Can I sit here in the lobby? Can I use that plug by the candy machine?" asked Cole, pointing to the electrical outlet by the bench lined with patrons.

"We don't serve food in the lobby," said the waitress, unmoved by Cole's plea.

"Here's my money," said Cole, forcing the 20-dollar bill into her hand. "I'll take a table when one's available. Okay?"

"I can bring you something to drink," offered the waitress.

"Sure, that'd be great," said Cole, heading for the outlet.

"What do you want?"

"Pardon me?"

"What do you want to drink?"

"Anything would be fine."

"Do you want a beer?"

"Sure. Anything," said Cole, busy plugging the AC adaptor into the outlet. He flipped the power switch. He squeezed into a spot on the crowded bench and strapped on the headset. The video monitor flickered and came on, filling the display with a dog's face. Cole jumped and banged his head on the overhead shelf. The café patrons paused mid-fork to watch Cole as he spoke, oblivious to his surroundings.

"Shoo! Scat!" he said, clapping his hands toward the diners. The dog sniffed around the hydrant. Cole spun the joystick and rammed the dog with the hydrant, making it yip. "Get out of here," he said and gave short chase to the mutt.

Evelynn was worried about her great-aunt. Dorothy Greene had called her and said to come over quickly, but when she arrived her aunt couldn't remember what was so important. Evelynn feared Aunt Dorothy was becoming demented. Then, not long after she'd arrived, Aunt Dorothy said she had an important appointment and escorted Evelynn to the door before she finished her tea.

Evelynn stood confused on the front porch of her aunt's house. Then she saw a fire hydrant attack a dog. The dog yipped and ran down the street. The hydrant chased the dog for a short distance, then stopped and looked directly at her. "Is this your car?" asked the hydrant.

Evelynn looked around then pointed to herself. "Are you talking to me?"

"I don't see anybody else standing with you."

Evelynn remained on the porch—silent.

"Is this your car?" repeated the hydrant.

"Yes…why?"

"You're parked in front of a fire hydrant. It's illegal to park in front of a fire hydrant."

"You weren't there when I parked. And besides, you're not real."

"I am real."

"I meant, you're not a *real* fire hydrant. You're a toy. And since there's no law against parking in front of toys, I'll be on my way."

"You want a mug?" asked the waitress.

"What?" asked Cole.

"You heard me," said Evelynn.

"Not you," said Cole.

"Not me' what?" demanded the waitress.

"A bottle is fine, thanks," said Cole.

"A bottle of what?" asked Evelynn. "Forget it. This is too weird. I'm leaving."

"No, wait. Don't leave," said Cole.

"I have other customers," said the waitress.

Wait…both of you," said Cole, pointing into the now silent restaurant.

"Both of us?" said both Evelynn and the waitress, both looking around.

"You can go. Thanks for the beer," said Cole, pointing vaguely to the crowded dining room. "But you must stay with your car."

"I don't have beer and I'm not staying," said Evelynn."Goodbye."

"Then you're under arrest."

"You can't arrest me. And for what?" asked Evelynn, pushing her fists into her hips.

"I can so arrest you. You parked by a fire hydrant."

"You are a toy," said Evelynn, nearly shouting.

"You didn't know I wasn't real when you parked here. So, your *intent* was to park by a fire hydrant. You are under arrest. Up against the car," demanded the hydrant.

"I will not. I'm leaving."

"Just a minute. I'm calling the police," said the hydrant.

"Right. And now what? Are you going to change into a phone booth?" scoffed Evelynn.

A sheriff's car screeched to a stop beside Evelynn's car. Officer Parker got out and approached the hydrant. "Did you call?"

"Yes, officer, I did," said the hydrant. "This woman parked in front of me. And she's resisting arrest."

"He wasn't there when I parked my car," argued Evelynn. "And besides, he-is-a-toy!" She gave Officer Parker a puzzled look. "Why's your hand in a cast?" she asked, pointing to his freshly plastered arm.

Officer Parker stared at his broken fingers protruding from his bright white cast and did not know what to say.

"That's what happened last time somebody resisted arrest," chirped the hydrant. "You should see the other guy."

"Well, ma'am. We had better discuss this at headquarters," said Officer Parker.

"Look, I wasn't parked here for long," pleaded Evelynn. "I just came over to visit my sick aunt."

The house door opened, and Dorothy Greene hustled down the steps.

"Aunt Dorothy," called Evelynn. "Wh...where are you going?"

Dorothy Greene paid little attention to the flashing red lights, the officer in the cast or the talking fire hydrant. "Oh, hello, deary. Have a nice time. I've got to go," she said as she bustled down the sidewalk.

"Was that your *sick* aunt?" asked Officer Parker.

"She didn't look sick to me," said the hydrant.

"So what? Are you going to arrest me because my aunt isn't sick?" asked Evelynn, her arms folded tight across her chest.

"You made a false police report. Don't you agree, officer?" quipped the hydrant.

Officer Parker pulled handcuffs from his utility belt. "We can go the easy way or the hard way, ma'am."

Evelynn could only retort with a disgusted huff. She shook Officer Parker's hand off her elbow as he escorted her to his patrol car. Officer Parker put Pluggy in the front seat with a seatbelt and Evelynn in the backseat behind the security screen.

The sharp turns and vibration of the drive made Cole car sick back in the café. He tried to pull the headset off, but the buckles wouldn't unclasp. He closed his eyes but kept the headset on because he didn't want to miss their final destination. He kept asking, "How much longer? Are we there yet? How much farther? Are we there yet?"

During the drive, Officer Parker reached for the box of donuts on the dash. He lifted the lid to the half-empty box and rifled through the remaining donuts. He gave a side glance at the rearview mirror and put the donut back, not wanting to binge in the presence of a lady.

They drove across town and made a sweeping turn through the towering arch of the Brookfield Industrial Complex.

"Where are you taking me?" asked Evelynn.

"Just a little stop first, ma'am."

They wove through the warehouses until they pulled in front of the Pioneer House.

"What are we doing here?" demanded Evelynn.

"You'll see, ma'am. Step out, please."

Officer Parker escorted Evelynn to the front door of the Pioneer House, then returned to the patrol car and drove away.

"I think our work here is done," said the smug fire hydrant. "Hungry? I know a charming café."

After Evelynn watched the patrol car disappear around the corner, she turned her attention to the front door of the Pioneer House. She tried the handle. It opened.

The dark interior of the Pioneer House was illuminated with a thousand candles. A lone table, graced with white linen and set for two, sat in the corner of the expansive room. In the porthole window of the kitchen door, she saw Aunt Dorothy and Lilly Hoffman jockeying for position, their faces pressed tight against the glass.

Terry Putnam stepped from the shadows. "Hello Evelynn," he said, then stared at his feet and blushed.

Startled and slightly confused, Evelynn demanded answers. "What's going on, Terry?"

"I wanted to apologize for the dinner we had at the firehouse. I wanted to make it up to you."

"Well, you certainly have a weird way of going about it."

Putnam stepped closer and softly kissed Evelynn's hand. "I'm sorry for everything. I'd like to start again. From the beginning. A new beginning."

Evelynn started to speak, but Putnam put his fingers gently to her lips, "Shhhh." Then, he delicately held her hand and tucked his arm around her waist.

In the far corner of the dance hall, Helga Broadenhoff wound the phonograph and lowered the needle onto the spinning album. Swirling to the crooning of Perry Como, Terry Putnam swooned Evelynn Dewitt with his three weeks of crash-course waltz lessons.

CHAPTER 18

RUN FROM HELL

Dennis Upman's head bobbed heavily, then he snapped erect. The reflective lane markers streaking through the headlights were his only orientation as the highway slipped into the peripheral darkness. The amphetamines he took three hours prior were losing their strength. He held the steering wheel with one hand and searched for the vial. His fingers fumbled over the sun visor and through his jacket pockets, but he found nothing. He frantically rifled the glove box, door pocket and under the seat. He went through his jacket pockets again, but his uppers were gone.

Rus Barker rubbed a chamois around the gauges on the pump panel, while sipping from a coffee mug hinged on his finger. The sun beamed through the coffee's steam turning it silver.

"Good morning, Rus," said Cole, as he stepped from the office after completing the morning's superfluous paperwork. He stretched his arms and yawned in the warm sunlight pouring through the open bay doors. "It's going to be a hot one today."

Cole stood in the bay opening and surveyed the weather conditions. The summer sun was already high in the sky. The city air hung dirty and stagnant; the asphalt wavered with radiant heat. Soon, the temperature will be unbearable.

Dennis Upman slapped his cheeks and drank the last of his tepid coffee. Clinging to the steering wheel of his 18-wheeler, he drove toward the rising sun. He had just picked up his load from the Torrance refinery. Licensed as a *Suicide Jockey*, he didn't give his cargo any extra thought. Bales of cotton or tanks of gasoline, money was money.

The morning sun filled the cab with soothing warmth. He shook off the chill of the evening ride. Bugs from the farmlands off the Interstate peppered the windshield. He tried the windshield wipers to clean away the bug guts, but the washer fluid ran out, leaving the glass streaked and hazy. Bright sunlight glared through the bug-streaked windshield, forcing him to squint.

This was his 37th hour of driving and he felt every minute of it. He knew several truckers who were lulled into sleep by highway hypnosis. The rhythmic hum of the drive, dreams of a soft bed and the embrace of a young siren seduced many a trucker into the arms of Morpheus.

"I've got to quit these killer drives," he said, shaking his head and slapping his cheeks, hoping to clear the cobwebs. He dug his hand through his jacket pockets and into the fold of the seat. He did not know what happened to the meth. What he did know was he needed to find a coffee shop. His thermos was empty, although the caffeine wasn't working as well as it used to. The uncomfortable fullness of his bladder was what kept him awake, but due to his bladder's state of capacity he decided to stop for relief while the decision was still his.

The walk to the diner didn't invigorate him like he had expected. His feet felt like lead. His mind was numb. Standing alone in the dirty bathroom, resting his head against the dirty wall, he closed his eyes for a moment. "Now, that feels good," he confessed to the urinal followed by a pleasant shiver. "Not as good as sex, but it's definitely a contender."

His boots were almost dragging as he shuffled back to his idling truck. Ahead was the drive to Norco, but after that delivery he promised himself some sleep. He hoped the drive through Los Angeles would keep him stimulated for the remainder of the haul. As he tugged on the steering wheel, the old Kenworth had no problem pulling the two shiny tankers and the 16,000 gallons of gasoline back onto the highway.

Kathryn Donaldson slipped from her muslin bath towel and slid her shapely legs into the cocoa nylons, then stood from the edge of the bed and wiggled into the tops of her panty hose. Today was not just another Tuesday but was the day of the promotional interview. This was the day that would change the rest of her life.

Not bad for a girl nearing 30, she thought, as she eyed herself in front of the wardrobe mirror. After a successful diet, she had rewarded herself with a modestly modest, coal gray business suit and skirt. It looked good on her.

Better than a bath towel. She admired her long, shapely legs that meandered up under her A-line skirt. Then, she pushed and patted her breasts, but nothing gave her cleavage.

She unrolled her hair from the curlers, allowing her thick auburn hair to bounce against her shoulders, then she leaned into the mirror and primped her make-up. She tried to guess the quality of lighting in the CEO's office. Too much light and not enough make-up would make her look too wholesome – too naïve; too inexperienced; too weak. On the other hand, too much make- up would be distasteful, implying poor judgment or a sleaze. It was a narrow line to walk.

"Ooooh!" she exclaimed as she looked at her watch. The interview was in an hour. She wanted to review her notes once more, but time was running out. Kathryn Donaldson took one last glance in the vanity mirror, then snatched the stack of index cards from the dresser and scampered out the door.

Engine 61 returned from the grocery store. A splash of air filled the apparatus room when Engineer Barker set the parking brake. He hooked his arm through the handles of the six plastic shopping bags and headed for the kitchen.

"I wonder what culinary masterpiece he's got planned for us today?" said Razmunson, as he wiped the chamois across the paramedic truck. Beads of perspiration collected on his brow.

"He only knows three summer recipes," said Valdez. "Fish tacos, pasta salad and cheeseburgers. It's too hot to barbecue, so the old fart will cook tacos."

"Why won't he make pasta salad?"

"He hates pasta salad," said Valdez.

"Well, whatever he makes, I hope it's cool," said Razmunson, wiping at the drips of sweat running down his cheeks. Valdez nodded in agreement as he resumed checking the equipment.

For trucker Dennis Upman, his biggest concern was staying awake. He forgot to get more coffee at the truck stop diner. He tried pinching the inside of his thigh to stay awake, but that trick only worked for the duration of the pinch. Besides, he didn't want to show up at his girlfriend's house with bruised thighs. It would cause too many questions. After all, she did have reasons to be suspicious – like the time she caught him with the waitress in Atlanta; and one in Dallas. His mind drifted for a moment as memories of his girlfriend played against the backdrop of his eyelids.

His head snapped back. He had fallen asleep! He hated when that happened. There was no warning. You're driving along and the next thing you know, you're waking up. He didn't remember falling asleep. He promised himself in just a few hours he would get some rest. He stared at the road and concentrated on staying awake.

His head snapped back again. He had fallen asleep again! He slapped his cheeks, then groped at the radio knob, delicately moving aside the dangling strands of *The Rolling Stones* draping from the mouth of the tape deck. He spun the volume dial to loud. His two 8,000-gallon trailers were riding level, though the tractor pulled a little to the left. He pushed his big rig east into the sun.

The gentle arc of the transition road put his truck on the 91 Freeway. From there, it was a straight shot to Norco. Traffic was light. To help stay awake, he busied himself reading billboards aloud. His favorites were rum posters. In particular, the one with the beautiful bikini clad, well-tanned female sitting on a secluded beach. It reminded him of the summer he and his ol' lady vacationed in Corpus Christi. He could still feel the warm sun and the gentle waves rocking their boat as the truck tires crossed over the raised line markers.

At least traffic is light, thought Kathryn Donaldson as she gave the rearview mirror a twist and studied her eyeliner. She steadied the steering wheel with her knee, while she fussed with an unruly eyebrow follicle. She gave the road a glance; it was clear. She returned the mirror to its correct position, then rummaged in her purse. Since there was still some time, she decided to rehearse her introduction. She knew her lines by heart, but kept the flashcards for reassurance. From her purse, she pulled the bundle of dog-eared index cards strapped together with a rubber band.

Checking the other lanes, she saw no other cars close, just a diesel truck in the right lane. She was glad traffic was light; she'd be embarrassed if somebody saw her talking to herself. She recited her introduction, her credentials and experience, then she drew a blank, forgetting what came next. Panicked, she looked to the center console at the stack of flashcards and flipped through them. The stack of cards slipped from her grasp and fell between her seat and the center console. "Darn!" she said and blindly groped beside the seat to retrieve the scattered cards. She pushed her arm deeper between her seat and the console, temporarily losing sight of the road. Quickly peeking over the dash, she froze in horror as her windshield filled with the broadside of a gasoline tanker.

At that exact instant, the dual wheels of the front trailer rolled over the hood of the little car. The truck driver stomped on the brake pedal, locking the wheels on the little car and skidding the truck into a jack-knife. The truck cab bounced violently as the momentum of the 50 tons of gasoline pushed the tractor sideways. Strapped in his seat, Dennis Upman watched helplessly from his side window as freeway traffic swerved to avoid his sideways truck. The rear trailer's tow bar broke loose and sliced into the front tanker, piercing the baffled chambers and bleeding the contents. The tractor cab shook and rattled as the right tire tore off its rim. The steel wheel dug into the cement freeway, releasing a piercing squeal and a flotilla of sparks showering the little car pinched between the jackknifed trailers.

From inside the little red car, Kathryn Donaldson's entire view through the windshield was the large truck tires on her smashed hood. The view from her left window was the front silver tanker; from the right window, the rear tanker. Her vision blurred from the violent vibration of the skidding, intertwined vehicles. She stiffened for the impact.

The truck skid crosswise over all four traffic lanes. The front tanker created an anvil for the back tanker to fulcrum against, pinching the little car. The right side of the car caved, vicing the console against Donaldson's arm. Her head snapped brutally to the left, slamming against the doorframe, then forward smashing her mouth against the steering wheel. The passenger window exploded, showering her with glass. The weight of the tanker trailer sank the dual tires into the hood of the little car, collapsing the dashboard onto her legs. Her right femur snapped, stabbing bone shards through her flesh and nylons. Small cracks webbed across the windshield, lofting a cloud of glass dust through the car.

The entwined vehicles banged hard against the cement embankment and rocked to a stop in the center three lanes of the freeway, near the Brookfield off-ramp. The delicate sound of broken glass and plastic car parts tinkled as they bounced down the freeway lanes. Kathryn Donaldson's battered head lay lifeless against the door, blood and saliva streaming from her mouth.

Dennis Upman kicked open the tweaked door of the tractor cab. The heels of his cowboy boots grated into the broken glass when he jumped from the rig. His mouth dropped open as he surveyed the carnage, then he ran up the off-ramp.

Kathryn Donaldson peered dazed through her veil of matted hair, unable to focus on the tanker pressing against her windshield. Her head shook with each heartbeat. She tried to scream but couldn't as the steering wheel pressed hard against her chest. She could not move her right hand and her legs hurt like hell. She fumbled at the chards of bone extruding from her thigh. She could smell gasoline. With her free left hand, she swiped at the clumps of bloody hair shrouding her face. To the left, from under the belly of the tanker, she saw a glimmer of flame.

Ding…

"Dang it," said Putnam, as he scooped a glob of beans off his breast pocket with a corn chip. "Why can't these people wait 'til after lunch?"

"It never fails," huffed Valdez, as he clanked his fork onto his plate. "They must broadcast some sort of public announcement when we sit down to eat."

Barker snatched his two tacos and headed for the engine.

Ddddrrrriiiinnnngggg!

"Engine Sixty-one, Squad Sixty-one. Traffic Collision. Eastbound Ninety-one Freeway at Brookfield. Twelve oh-two."

The engine and squad responded down the wide double lanes of Main Street. Ahead, Cole could see cars sitting idle on the freeway overpass. When they were at the entrance of the off-ramp, he saw a small black column of smoke swirling into the sweltering summer sun.

"Something's burning, Cap," said Barker, pointing at the smoke with his taco.

"A lot of something," added Cole. As they crested the on-ramp, they saw—over a quarter mile away—a jackknifed tanker truck blocking all lanes of traffic. The main body of smoke was coming from the front tanker. The oblique angle of the rear tanker blocked most of their view of the tractor cab.

"We're going to need some help," said Cole. He pulled the radio mic close to his mouth and spoke over the siren's wail. "L.A. Dispatch. Engine Sixty-one. We have a jackknifed fuel tanker involved in fire. Start two more engines, a ladder truck with the jaws and also notify CHP. We'll need traffic control."

In a calm, professional tone, the female dispatcher repeated his order. The equanimity of her voice was soothing.

Engine 61 edged cautiously along the freeway shoulder to the right of the congested traffic. Their siren confused the motorists, causing the cars to move in various directions. Nearing the accident, they saw reddish-orange flames raging beneath the tanker. Shards of flames leaped above the tanker, disappearing into the seething funnel of smoke.

Cole knew there was inherent difficulty getting water to a freeway fire, especially one of this magnitude. There were no fire hydrants on the freeway. The 500 gallons of water they carried on the fire engine would last only four or five minutes with their smaller handlines, or less than a minute if they used their water cannon.

"Stay to the side of the tanker, Rus," ordered Cole. He knew the most dangerous sections of the tanker were the ends. If the tanker BLEVE'd, it would most likely tear in half and launch both ends like rockets. Anything positioned at the end of the tankers was like being perched at the barrel of a very large cannon. A slow line of cars snaked to the left of the tanker and along the right shoulder to exit from the Brookfield off-ramp, but the main body of traffic was gridlocked about 50 feet from the burning tanker. The shiny end of the tanker sat cantered, pointing lethally at the stalled traffic. Some of the commuters were standing out of their cars, hoping for a better look down the flight path of the 20-ton rocket.

Cole studied the predicament cautiously. Rushing in too quickly and establishing an inappropriate position could put the entire crew in jeopardy. It was safest to size-up the situation before committing to a position. He opened the engine door and stood on the running board. Although the rear tanker still blocked their view of the truck cab, it was a better vantage point. Cole could see nothing more than the two tankers and the top of the tractor. The DOT placard on the tanker read *1203*. Cole recognized the number from his experience in the Petroleum Storage Bureau. It was gasoline.

The first tanker was ruptured by the rear trailer's tow bar. Burning fuel issued from the incision and flowed under the second tanker, which was still intact and full of gasoline. Figuring they would have to check the cab for occupants and cool the tankers, Cole ordered Barker to position the fire engine on the uphill side of the tanker at the mouth of the off-ramp. As the fire engine edged around the tanker, the red sedan came into view. Wedged in the nadir of the two tankers was a small red car. Flames were impinging on the hood. From his elevated position on the fire engine, Cole could see the woman inside.

"Pull a transverse line!" Cole shouted to Putnam. "Keep the fire away from the driver!"

Barker swung the fire engine into position, then jumped from the cab and slid the transmission lever into *pump*. Putnam pulled the hose from the center hose bed and ran toward the fire. The fire hose sizzled against the pavement as Putnam dragged it across the freeway lanes. He hid behind the rear of the little car; it was too hot to get any closer. "Water!" Putnam yelled from his crouched position, raising both arms above his head. "Water!"

Barker acknowledged Putnam's arm signal and pulled the handle to the discharge gate. A bolus of water swelled the hose at the discharge connection, then the bulge snaked down the 150-foot length of hose. The lump in the hose picked up speed when Putnam opened the nozzle. Trapped air hissed as it blew from the nozzle tip. Putnam turned his attention to the fire. Huddled at the rear of the crushed little car, he was well between the ends of both tankers. He tipped his helmet brim down to shield his face from the radiant heat. The woman trapped in the car was desperately squirming to free herself from the press of the steering wheel and to hide her face from the searing flames.

The nozzle jerked and coughed as a combination of air and water sputtered out before changing into a solid stream. The water exploded into steam as it impacted the flames in front of the car. Beads of water danced on into steam as it impacted the flames in front of the car. Beads of water danced on the car hood as they boiled into vapor. Putnam rotated the nozzle tip changing the straight water stream into a wide protective fog pattern. Cool air rushed past him as it was sucked into the venturi of the spray. He advanced into the concave prow of the wreckage. He continued into the tight corridor between the tanker and the car until water sprayed past the driver's door, putting a thin shield of cool water between the woman and the inferno.

From his position near the side of the driver's door, Putnam studied the woman slumped over the steering wheel. Her face was obscured by her hair, tossed and matted with blood. She was mouthing words that he could not decipher; only moans and gibberish that may have been prayers.

Valdez came behind Putnam and nudged him with the drug box. Putnam twisted sideways, but Valdez could not squeeze past. There was not enough room for both of them between the car and the ominous tanker. Putnam backed up, inadvertently angling the water spray away from the fire. Angry flames of pressurized gasoline shot from the tanker's incision burning with the roar of a jet engine. A swath of fire leaped across the hood. Both firemen flinched from the heat. Loose pieces of chrome trim clattered as the little car vibrated from the rage of the fire. Putnam whipped the water spray back into position, forcing the flames into check.

Valdez edged forward, walking sideways in the narrow passage. He shouted directly into Putnam's ear, "How is she?" He turned his ear to Putnam encouraging a reply.

"Don't know," yelled Putnam, using short words so he could be easily understood. "I saw her move," he said, shrugging his shoulders.

There was even less room between the driver's door and the tanker shell. The dual wheels of the tanker sank into the hood of the sedan, pressing it low to the ground and collapsing the passenger space. The gouged and scratched shell of the gasoline tanker rested ominously against the windshield. The car roof was buckled inward, further decreasing the car's interior space.

Valdez knocked on the driver's window, the only one left unbroken. His thick leather glove made a dull thud against the noise of the fire. The woman slightly moved her hanging head. Her face was obscured with straggly hair. He could see her lips move, but she did not make any conscious effort to communicate. Searching in the pocket of his turnout coat, he pulled out a spring-loaded center punch and pressed it against the window. The punch *clicked*, instantaneously shattering the window into thousands of small cubes of glass.

Abruptly exposed to the heat and the fire's roar, the woman came alive screaming and flailing her left hand mindlessly. Valdez captured her thrashing arm. A quick check of her pulse found it weak and rapid, a sign that she was hemorrhaging and going into shock. He stuck his head inside the car, but his helmet hit the collapsed roof. He pulled off his helmet and leaned through the window. The inside of the car was hot and dank, and wreaked of strawberry air freshener and radiator fluid. He was almost nose-to-nose with the woman. She was panting. Her eyes hysterically darted back and forth, not stopping to focus on him.

"Ma'am... Ma'am, can you talk to me?"

The woman did not respond. Her lips quivered. Her head flopped from side to side. Valdez gently pulled on her right arm. She shrieked into his ear. He palpated her mangled legs. She screamed again. The compound fracture had pierced her femoral artery. A fine mist of blood pulsated into the air. He pulled a knife from his pocket and cut the seat belt, then tied it around her thigh as a tourniquet. The woman screamed relentlessly, but the bleeding stopped.

Valdez leaned back from the car and saw Razmunson standing at the back bumper, staring blankly at the tankers. His face was pale and moist.

"Raz...Raz!" shouted Valdez.

Putnam nudged Razmunson out of his trance.

"Throw me a tourniquet. Set up two large-bore IV's and tell the Cap we need the jaws."

With a nod, Razmunson assembled the equipment and gave it to Valdez, then ran to the fire engine and relayed the information to Captain Walker.Cole had anticipated the extrication and had already called for Truck 45.

Boom!

The little red car shook violently as one of the tractor tires exploded. All the firemen dropped to the ground. Large sections of flaming tire tread sailed over the fire engine. The rear tanker rocked back, pulling the trailer yoke from the hole, releasing more fuel to the fire. Putnam felt the warmth of a small release of urine.

Engineer Barker had a small handline that he sprayed on the fire engine. The plastic lens covers had already melted from the heat. The cool water spider-cracked the right side of the windshield and passenger window. Barker also felt the warm sensation of urine following the tire's explosion.

Cole ran to the tractor cab and pulled the clipboard from the pocket on the driver's door. The bill of laden confirmed his fears. The two tankers' total capacity was 16,000 gallons of gasoline and each tank was filled to 85 percent capacity. He knew the tanks were most vulnerable at the top—the empty part. With no liquid to cool the heated tanker shell, the metal would eventually fatigue from the pressure of the boiling fuel and BLEVE.

"Engine Sixty-one, L.A.," called a female dispatcher over the radio.

"Sixty-one, go," Cole yelled into his radio.

"Engine Sixty-one be advised, Engine Forty-five is delayed by a train. Next-in is Engine Thirty with a four-minute E.T.A."

Cole ran back to the fire engine. "How much water do we have left?" he shouted at Barker. The rumble of the engine's pump added to the deafening fire-ground noise and Bleve, poised on the engine cowling, barked at the burning tanker further adding to the bedlam. Cole pulled Barker by the sleeve and shouted directly into his ear. "How much water?"

Barker tapped the sight gauge on the instrument panel. "About a hundred gallons," he shouted into Cole's ear. "One minute," Barker added, rising one finger.

"Gate the line down. Our water supply is delayed. We need four more minutes."

Barker scowled as he throttled back the engine pump and pushed in the discharge gate handle.

Cole scanned the horizon for Engine 30. He contacted them by radio and ordered them to lay a supply line from the closest hydrant, which was 200 feet past the end of the Brookfield off-ramp. Two offramps away, Cole saw Engine 30 racing up the frontage road.

Cole took a moment to survey the area. He saw the overpass was already lined with spectators and television crews, while five helicopters circled overhead. The Highway Patrol had closed the freeway in both directions and was directing the stranded motorists to back down to the next off-ramp. They had the freeway shut down a half mile in both directions. Cole wondered if that was far enough.

The hollow clacking of hose couplings captured Cole's attention. Engine 30 was racing down the off-ramp as four-inch hose paid off their hose bed. Engine 30 was at the bottom of the off-ramp when the last hose coupling paid out and landed with a hollow clank. They ran out of hose 100 feet short of Engine 61.

Barker pulled three sections of hose from Engine 61's hose bed and tossed one end to Cole. While Cole connected the hose lines together, Engine 30 raced up the off-ramp to pump the supply line.

"How much water is left?" Cole shouted over Barker's shoulder as he spun the hose couplings onto the engine's intake valve.

"Less than 20 gallons," said Barker after a quick look at the sight gauge. "The pump will cavitate in about 20 seconds."

Cole looked up the off-ramp searching for Engine 30. He saw the engine at the hydrant down Brookfield Boulevard. He shouted into his radio mic. "Engine Thirty. Engine Sixty-one. We need water now!"

"Copy, Engine Sixty-one. It's on its way."

Cole strained his eyes, watching the flat hose bulge with the telltale sign of incoming water. He stepped on the hose. It was soft and spongy from the residual air that preceded the water.

Air hissed from the small air vent at the pump intake. The bypass vent released trapped air before the pump was voided of water. The small vent whistled as it tried to bypass the large volume of air trapped in the thousand- foot supply line. The fire engine's diesel motor madly raced as it free-spun the air-filled pump. Barker slammed the throttle panic button before the pump self-destructed.

Putnam stumbled forward when the stiff hose he was leaning against fell limp. Flames leaped from under the tanker, engulfing it from both sides. Putnam scrambled behind the car. Valdez dove inside the driver's window to escape the heat. His body and Nomex turnouts shielded the woman. The back of his legs, exposed through the window, smoldered from the intense heat. Razmunson ran for the protection of the fire engine.

Cole and Barker stood frozen as their attention was diverted. from the shrill whistle of the engine's by-pass vent to the deep, throaty bellow of the rear tanker. On top of the tanker, the lid to the pressure relief valve slapped open and spewed flames. A two-foot column of pressurized, angry flame ejected from the eight-inch vent. A loose trailer fender rattled defiantly with the jetting flames. The car's console vibrated against Valdez's chest and coins in the ashtray rattled as the little car shook from the quaking tanker.

"Do something!" yelled Valdez, his face buried in the contorted cushion of the passenger seat. "Cool me down!"

"Water!" shouted Putnam from his crouched stance behind the car. The air was hot. He shielded his face with his helmet and took quick, shallow breaths from behind the collar of his turnout coat. He looked back at the fire engine; there was no one in sight. Even Bleve had retreated from the engine cowling to the shelter of the left jump seat.

Air continued to scream from the pump's small intake by-pass. Barker stood stunned watching the distant bulge in the supply line slowly descending the off-ramp.

"Open the gates!" yelled Cole, but Barker was fixated on the lazily approaching water.

"Open the other gates!" he shouted again, with no response from Barker.

Cole pushed Barker aside and pulled the T-handles to all of the engine's discharge gates. Cool air blasted from the discharge openings. The bulge in the supply hose raced down the off-ramp to the fire engine. A heavy mist sprayed from the open discharge ports just before they burped and spewed water. Barker took over the control panel, shutting the open-butt gates and throttling up Engine 61's main pump.

The limp fire hose draped across Putnam's arm stiffened. He advanced against the fire as the nozzle coughed and sputtered into a spray of water. The fan of cool water splashed across Valdez's legs and the side of the woman's head sticking her straggly hair to her face. Putnam pressed forward, forcing the flames to retreat beneath the tanker.

"Shit!" said Valdez, pulling himself from the car window. Nobody heard him over the tanker's bellowing relief valve. "That thing's gonna blow!"

Bang! A loud explosion shook the firemen, and the little car was engulfed in a stinging cloud of steam. The tanker moaned and buckled as Engine 61's water cannon hammered into its side. Perched atop the engine, Razmunson played the heavy water stream across the tankers. As the rear tanker cooled, the flame jetting from the relief valve dwindled until it was scarcely visible and its moan imperceptible. Barker relieved Razmunson at the handle of the water cannon. Captain Walker trotted to the other side of the tanker to survey its condition. Bleve returned to his position on the engine cowling and resumed barking at the tanker.

On the other side of the tanker, Cole noted the trailer yoke had pulled further from the front tanker. Fuel gushed unrestricted from the breach. Burning gasoline flowed underneath the second tanker and long flames impinged directly against the vulnerable upper shell. The top of the tanker had turned a purple and black hue. Fatigue and failure were imminent.

"Engine Forty-five, Walker. Take position across the freeway and cool the rear tanker with your water cannon." Engine 45 acknowledged his order.

Heavy droplets of water pounded like fists against the top of the little car. A thick stench of petroleum hung in the humid air. Valdez leaned further through the car window to examine the woman and escape the pelting water. He replaced the seatbelt around her thigh with a restricting band and established an IV in her left forearm. He taped the IV line to her arm, the liter bag of saline clenched in his teeth.

At the other side of the car, Razmunson crept to the window. He tried to open the door, but it was wedged against the tanker. He pressed his face between the window column and the tanker but jerked back because the vibration blurred his vision. "What do you need, Val?" he shouted through the broken rear window.

"A vacation day," Valdez replied as he slid back out the window. The woman grabbed his collar as his face passed hers. She pulled him to her until the tip of her bloody nose pressed against his cheek. Though her eyes were half-closed and glazed, she spoke with unexpected clarity.

"Don't leave me here," she begged. "Please don't let me die," she said before sobbing. Her mouth opened and her lips quivered, though no sound came out.

Valdez pulled her fist from his collar. "We're working on it."

Under the rain of steamy, oily water, Valdez joined Razmunson and Captain Walker at the rear of the car.

"Where the hell's the jaws?" Valdez shouted at Captain Walker.

"Truck Forty-five is one minute out. How's the lady?"

"She's screwed up, Cap. She has a broken femur, a lacerated femoral artery, facial trauma, a flailed chest, her foot is pinned under the brake pedal, her right hand is crushed between the seats, and she probably has abdominal damage. In short, she's circling the drain." Valdez shook his head in disgust and climbed through the back window to his patient.

Truck 45 sped down the off-ramp, swung a wide arc and pulled parallel to Engine 61. Bleve turned from barking at the burning tanker to barking at Captain Decker. Cole grabbed Bleve, put him inside the fire engine cab and closed the door.

Holding up his hand, Captain Decker held his crew in position on the ladder truck then sauntered to Captain Walker. During his approach, he studied the flaming tanker almost like he was absorbing its energy and challenging its bluster. He put his hand on Cole's shoulder and shouted into his ear. "What do you need?"

"We need the jaws. The driver is wedged tight. We'll need the roof cut off and the woman extricated."

Captain Decker gave Cole a nod and made a couple of gestures with his arms, sending the three truck firemen racing to various equipment compartments. Two of the firemen carried the hydraulic power unit, while the remaining fireman carried the spreader and the shears. The truck firemen connected the hydraulic lines to the spreader and fired up the power unit within seconds. The four-stroke motor cycled and loped as they brought the power tools to the little car. But the large truck firemen could not fit between the tanker and the car. There was simply not enough room. Valdez laid the IV bag on the dashboard and yanked the shears from the baffled truck fireman.

Valdez placed the cutting shears around the car's back roof columns. As he twisted the control handle, the hydraulic shears sliced effortlessly through metal, severing the left rear post in seconds. He slid across the trunk of the car, slammed the hardened steel blades through the remnants of the back window and cut the right rear roof support. The truck firemen stood on the trunk lid and pressed the roof up and forward toward the tanker. Seething water showered the woman, intensifying her screams. Eventually, her mannerism eroded from sheer terror, to modest complaining, to complacency, to consignment, to prayer.

Boom!

The ground shook and both tankers clattered as two of the rear trailer tires exploded, sending lengths of flaming retread through the air. The rear tanker rocked backward, loosening the trailer yoke from the front tanker's incision and releasing more gasoline. Flames leapt from under the tanker. Fire shot four feet high from the relief valve. The deep bellow returned. The woman screamed and frantically shook the steering wheel, trying to free herself from her steel trap. Then, just as quickly, she leaned back against her car seat and muttered what sounded like a Psalm.

Captain Walker scanned across the freeway. Engine 45 was setting up their water cannon on the frontage road directly across from the tanker. In moments, they would douse the opposite side of the tanker with cooling water. Inside the opened car, in pelting rain of oily water, Valdez straddled the woman and severed the steering column with the hydraulic shears.

Cole pulled Captain Decker's arm and leaned into his ear. "There's nothing for your crew to do. We'll handle the extrication. You should leave."

Captain Decker stared into Cole and started to reply, but Cole cut him off. "Please…go." His voice cracked.

Captain Decker's stern expression melted into one of respect. Captain Decker shook Cole's hand with a hard squeeze. He held his grip and pulled himself closer to Cole. "See you on top," he said. He then whistled and waved his crew back to the truck. The truck firemen sprinted to their positions. Captain Decker sauntered back to the cab of the ladder truck.

Before taking his seat, he stood on the running board and saluted Captain Walker. He plunked himself in his seat and pointed forward, signaling the engineer to drive. Truck 45 drove up the off-ramp, leaving the crew of Station 61 alone with the woman and the fire.

Lined across the distant overpass, news crews aimed their telescopic lenses at the inferno. Jennifer Preene, lead reporter for Channel 3, dabbed her nose with facial powder and primped her hair before giving the cameraman a nod. She counted to three and spoke into her microphone.

"Tragedy struck the peaceful town of Brookfield today when a tractor trailer carrying 16,000 gallons of unleaded gasoline collided with a small sedan, trapping the driver. With us is Los Angeles Fire Department's Public Information Officer, Red Hall. Officer Hall, what circumstances lead up to this terrible accident?"

The camera panned back, revealing Red Hall standing next to the reporter. She extended the microphone toward him, holding it tight, expecting Red to grab it. He didn't. "At this point in time, Jennifer, it is unclear exactly what caused the accident. What we do know is at a little after twelve o'clock this afternoon, Los Angeles Fire Engine Sixty-one was dispatched to a simple vehicle accident on the Ninety- one Freeway at Brookfield Boulevard. When they arrived, they found a burning gasoline tanker and a woman trapped in the wreckage. Besides being pinned in her car, her medical condition is critical."

"Officer Hall, tell us about the perils faced by the firefighters on this incident." The television camera panned beyond the reporter and zoomed to the firefighters miniaturized by the goliath flames. The commentary of Red Hall voiced-over the dramatic scene.

"The firefighters are ensnarled in a very precarious situation. They must extricate the trapped driver and endure the extreme conditions of the fire. The radiant heat can be over 1,200 degrees, and smoke coming from the burning hydrocarbons is lethal. Plus, in addition to protecting themselves and the victim in the car, they must keep the fuel tankers cool, so the tanks don't over-pressurize and explode...known as a BLEVE." Red Hall continued to explain Boiling Liquid Expanding Vapor Explosion to the reporter. He detailed the progression of events he was witnessing. By his own word, he realized the jeopardy of his friend and the firemen of Station 61. As he spoke, Red Hall turned toward the fire. He also was a helpless bystander, witnessing his friend and colleagues in a struggle for their lives.

Jennifer Preene interrupted. "How does it feel to watch your brother firefighters in a battle for their lives?"

Red Hall burned his stare into the reporter but turned away in silence. He pressed hard against the guardrail, unwilling to share his emotions with the media.

The camera panned from the tanker fire and refocused on the beautiful, placid reporter. From her safe vantage point, she began her summary. Suddenly, the microphone was pulled from her hand. In stepped Red Hall.

"I *would* like to tell you what it feels like watching my brothers battle for their lives." He pointed to the fire and told the cameraman to focus on the firemen. Jennifer Preene shrugged her shoulders to the man behind the camera. He focused on the fire.

"This is real, ladies and gentlemen. This isn't a sound stage. This isn't a soap opera or game show. Look at those men. They are real people, with real lives, with real families and real fears. They don't want to be there. They do what they do because the person in the car is a real person, too. And she needs help. So, put down your fork and push away your TV tray and pray for the firemen because they are all that stands between us and an ugly fate." Red Hall politely returned the microphone to the speechless reporter and returned to his helpless position against the overpass rail.

Removing the steering wheel had given the woman more room to expand her lungs, directly increasing the volume and length of her screams. Teetering next to her on the accordioned seat, Valdez wedged the hydraulic spreader between her seat and the center console, careful not to pinch her mangled arm. He twisted the control handle, deftly delivering 10,000 pounds of spreading force against the seat bracket. The gap between the seat and console widened. She tugged desperately on her broken arm. The gap widened further. She slid her limp, angulated wrist from the grip of the seat.

Standing ready on the back truck lid, Razmunson exchanged the jaws spreader for an arm splint and a roll of tape. The woman shrieked when Valdez scooped her fractured arm into the trough of the splint. He wrapped the entire roll of tape around the cardboard splint, hoping to protect it from the water deluge.

Directly across the freeway, the fireman from Engine 45 spun the worm drive of the 1,000-gpm deck gun and took aim on the tanker. The engineer throttled up the diesel motor, sending an arc of water across the freeway lanes. At full pump pressure, the water stream hammered against the side of the tanker, driving the fuel and flames underneath with explosive fury.

Instantaneously, Putnam's fragile water spray exploded backward, steaming his face. He fell to his knees but held his position on the fire. Flames clawed from under the tanker, searing across the windshield of the little car, filling it with choking, burning steam. Valdez dove to the floor pan. Huddled beside the driver door, Putnam's small handline could scarcely shield the woman from the blowing flames.

"Engine Forty-five. Shut down! Shut down! Shut down!" screamed Cole into his radio. "Shut down! You're burning us!"

Engine 45's engineer, not waiting for his captain's orders, slammed the throttle panic button, dropping the heavy stream from the side of the tanker. Putnam advanced his hose line, forcing the flames back beyond the car. Valdez slid out from under the dash and resumed scooping water from the floor pan with his helmet to visualize the woman's trapped ankle.

Cole paced between the fire engine and the little car, studying the tankers for signs of fatigue. The flame blowing from the relief valve stood six feet into the air. The deep moan had changed pitch, mimicking a scream.

"Val, how much longer?" Cole yelled from the back of the car.

Valdez shook his head without an answer, frustrated by the severity of the woman's entrapment and the cramped working space. The brake pedal pressed hard against her flesh, bending her right leg nearly in half. There was no room between the brake pedal and her foot for his fingers, much less the bulky shears. He shook the pedal with his fist. It did not budge.

"How much longer?" demanded Captain Walker.

"I don't know!" Valdez shouted, dropping the shears into the water of the floor pan. He climbed over the seats and jumped over the trunk, landing directly in front of Captain Walker. "Cap, this whole situation is out of control and turning to shit," he shouted. "Half her bones are broken, and she's lost a lot of blood. And I can't get her out!" Valdez was not shouting to overcome the roar of the fire; he was shouting because he was frustrated and scared.

Boom!

The tire on the hood of the car exploded, slamming the car's roof down over the woman. The rear tanker rocked violently, slicing the pintle hitch across the gored tank widening the gash. The woman screamed hysterically, adding to the tumultuous roar of the fire.

The firemen rose from the ground. Valdez remained hunched, holding his hand on his helmet. Cole pulled Valdez to his feet. Valdez met him with a hostile scowl.

"This thing is going to BLEVE, Cap. It's not going to be much longer. The lady is probably going to die right in the middle of this anyway. So, let's just leave and let this thing blow."

"No," said Captain Walker. "We are going to stay here and get her out."

Valdez jabbed his finger hard into Cole's chest. "Are you risking getting us killed for a body recovery, or television airtime, or what? Don't kill the rest of us because *you* have something to prove."

Cole stepped forward, closing the space between himself and his paramedic. "Val, the tanker is not going to explode…not yet. I know these containers. We still have some time. Trust me."

The angry fireman held his glare on Cole, digesting his words. He did not move.

"Val, we can't quit. Not now. We're too close." Cole put his hand on Valdez's shoulder and gave him an encouraging squeeze. "Let's free that woman and get the hell out of here." He gave Valdez another squeeze and gently turned him back to the car. Valdez hopped back onto the trunk and gave one final look at Cole before ducking back into the twisted wreckage.

Cole took a deep breath and slowly exhaled, hoping he was right.

The exploding tire forced Red Hall from the guardrail. He couldn't watch any more. He leaned back against the rail and watched Jennifer Preene interview Mayor Jenkins with his nauseating preamble on public servants and community spirit and dwindling tax dollars. Behind the mayor, out of the camera's view, stood Chief Pierce waiting for his 15-minutes of fame. Chief Pierce was pragmatic. This accident was an excellent public relations piece. If the firefighters were successful, they would be heroes and the grateful Board of Supervisors would grant lavish financial rewards to the fire department. If the firefighters were lost in their effort, they would be martyrs and the mournful Board of Supervisors would grant posthumous financial rewards to the fire department. Either way, the fire department would reap financial benefit. For Chief Pierce, it was a win-win situation.

A flash from the mayor's lapel caught Red's eye. He stood beside Chief Pierce and waved his arms for Jennifer Preene's attention. "Ask about the lapel pin," he mouthed, shaking his lapel. "The button. Ask about the button."

Jennifer Preene interrupted Mayor Jenkins mid-sentence as she studied the campaign button he wore. "Mayor Jenkins, what does your lapel pin mean, *Fire L.A. Fire?*"

Mayor Jenkins grabbed his lapel like he'd been stabbed in the heart. He clutched at the button and began to unpin it but stopped. "Oh, this…" he said with an uncomfortable, political laugh as he tried to explain his position by citing damaged coffee tables and broken punch bowls.

Red ran to his sedan and pulled out the Brookfield file folder and dumped the papers out. He ran back to Chief Pierce. "Can I borrow your pen?" he asked as he snatched the pen from the Chief's shirt pocket and scribbled across the folder. He flashed his scrawl at the reporter: *L.A. Fire saved his grandson.*

"Is it true that the Los Angeles Fire Department saved your grandson's life?"

Mayor Jenkins stood silent. Excuses log-jammed in his mind.

Red flashed the other side of the folder. *Face and throat burned.*

"How badly was his face and throat burned?" she pried.

Mayor Jenkins stammered for an answer as Chief Pierce stepped into the camera pan and put his arm around the mayor. "All in a day's work, right, Merle?" said the Chief. "The firemen were just doing their job. In fact…" said Chief Pierce, pulling the mayor in tight and turning him into the camera. "Mayor Jenkins is extending our contract. Right, Mayor?"

A hard rain of oily water fell on the firemen as the heavy water stream from the engine's deck gun passed overhead and hammered into the tankers. Only the soles of Valdez's boots showed over the passenger seat. He groped in the murky water of the floor pan, searching for a leverage point for the jaws. The woman screamed each time he bumped her mutilated leg.

Cole took a quick lap around the tankers. On the far side, unchallenged flames lapped the side of the rear tanker. The column of burning vapor ejecting from the relief valve shot over ten feet high and whistled like a boiling tea kettle.

Sweat and oily water streamed off the tip of Valdez's nose as he pounded the jaw's tip through the floor pan to drain the water. Her foot came into full view.

"Ma'am, this may hurt a little," he muttered as he wedged the spreader tips between the brake pedal and the raw tissue of the woman's ankle. With one tip against the pedal and one tip wedged against the transmission housing, he twisted the control handle, spreading the jaws. As the spreader tips widened, the massive tool housing fulcrumed against the woman's angulated leg. She screamed in agony and clawed at Valdez's face. She paused to inhale and shrieked louder as the spreading jaws bowed her fractured leg. The woman inhaled again and screamed louder yet. She inhaled again and cried out in anguish, pleading for Valdez to stop, then she fell silent.

Valdez glanced over his shoulder at the slouched woman and quickly checked her pulse. She had fainted much to his relief. He took advantage of her unconsciousness and powered the jaws open until the shaft of the brake pedal snapped. He tugged on her pulverized ankle pulling it free. Her leg felt like a bag of pudding as he lifted it from the twisted metal.

"It's loose," he said, startling himself. "She's free!" he shouted, waving for Razmunson.

"She's free!" he shouted over the back seat to Razmunson, who was fixated on the 20-foot flame, jetting from the relief valve atop the tanker. "She's free!" he shouted again and threw the woman's shoe, bouncing it off Razmunson's helmet.

Razmunson scrambled onto the back deck of the car and laid the Miller board across the seats. They pulled her small, limp frame onto the board. They quickly synched some of the Velcro straps and carried her from the car.

Captain Walker called for the waiting ambulance. The ambulance responded down the off-ramp, its siren faintly audible over the noise of the fire. It sided next to Engine 61, sheltering from the heat. The paramedics ran with the backboard, the woman's flaccid body quivered with each jarring step. They slid her into the ambulance.

"Go! Go! Go! Go!" shouted Cole, slapping the side of the ambulance.

Valdez jumped into the ambulance and barely closed the double doors as they sped up the off-ramp.

Razmunson started the paramedic truck and jockeyed it around the fire engine and over the four-inch supply line.

Barker prepared Engine 61 for their departure by shutting down the water cannon. The instant the water stream stopped, the shrill of the relief valve rose beyond human perception.

"Turn it on! Turn it back on!" ordered Cole, staring at the 30-foot column of flame.

"Drop the line, Terry. We're leaving it here," shouted Cole. "Jump on the squad!"

Scrambling for the accelerating paramedic truck, the three firemen climbed onto the rear step bumper. Cole banged on the utility box with his fist, signaling Razmunson to accelerate. "Go, go, go!"

As the paramedic truck sped away, Cole watched the tankers shrink in the distance, the vapor flame climbed steadily higher. "Faster!" he yelled, banging his fist on the top of the truck. There was no longer any audible sound from the relief valve. There was no noise other than the rumble of the churning fire and a muffled howl.

"Bleve!" Cole exclaimed as he jumped from the truck bumper. His feet could not match the speed of the pavement as he tumbled to the ground. Bouncing to his feet, he ran back to the inferno. The down slope of the off- ramp gave him speed. His heavy steel-shank, water-soaked boots slapped hard against the pavement, pulling him downhill. His vision bounced between the 40-foot flame shooting from the tanker's relief valve and Bleve's face framed in the window of the fire engine.

The uncontrollable speed of his downhill sprint slammed him hard against the side of the fire engine. He pulled down on the handle and yanked the door open. Bleve jumped into his arms. He tossed the dog toward the off- ramp. "Run, Bleve, run!" The dog bolted toward the paramedic truck cresting the top of the off-ramp. His toenails scraped against the pavement for traction as he galloped toward safety.

Spent from his sprint to the fire engine, Cole turned to run back up the off-ramp. His boots skidded heavy against the asphalt. He was already out of breath. His turnout coat hung weighty on his shoulders, laden with water and perspiration. Sweat poured from under his helmet, stinging his eyes. His chest heaved and his vision blurred as his safety zone remained a bouncing dot on the distant horizon.

Cole glanced over his shoulder and saw the water cannon no longer played across the tanker's side. Left unattended, the heavy stream had migrated off target and sprayed futilely between the two tankers. Dark orange flames and black smoke engulfed the majority of the tankers. The column of flame jetting from the relief valve created a vortex that sucked the black smoke from around the tanker and blew it skyward. Suddenly, for a fraction of a second, the towering flame stopped. No longer driven by pent vapor pressure, the venting flame succumbed to the tanker's shell failure.

The sequence of events following the tanker's release of its contents took seven-tenths of one second. As the tanker's shell tore along the weakened seam, over- pressurized gasoline sprayed through the tear, nebulizing over the fire. The fine fuel particulates ignited in a chain reaction, drawing the flame back inside the tanker. The introduction of the fire into the confined space increased the pressure exponentially, tearing the tanker completely in half with the ease of an aluminum beer can.

The back half of the rear tanker rocketed down the freeway, cartwheeling end over end, spewing a sphere of fire that engulfed all eight lanes of the empty freeway. The front half of the rear tanker thrust into the back of the front tanker like a piston, shredding it and spraying its lethal contents into a titanic fireball, consuming the water streams, hose lines and Engine 61.

Broadsided by the force of the BLEVE, Engine 61 rolled and tumbled, deflecting the massive fireball upward. The trailer's rear axle blasted down against the pavement, then rebounded skyward, over the burning fire engine and spiraled toward the off-ramp.

Committed to his run from hell, Cole was in a dead stride, but was still too far from the crest of the off-ramp when the tanker BLEVE'd. The shockwave reached him first. The compression impulse drove him headlong against the freeway pavement. With no time to break his fall, he skidded on his chest and face. His nose contorted and bent as it bounced along the pavement.

The time Cole lay stunned on the pavement saved him from the ensuing heat wave. Its scorching ellipse covered the entire area, flashing over him. The sharp burning fingers of the fireball dug through his turnout coat and seared his back. Instinctively, reflexively, he staggered to his feet and resumed his run.

The air was hot and denuded of oxygen; the flashover had consumed most of it. It was difficult to breathe. His heaving chest was sore, and each inhalation stole precious energy. He stumbled forward, gasping for air, as the mammoth flaming axle bore down on him. Flames clinging to the loose tire tread flapped in the air like unfurled flags. A three-foot length of shredded rubber extended from the burning tire whipped the air and spit a pinwheel of flames with each revolution. The gyrating three-ton axle spiraled down just above the shoulders of Cole Walker.

The hourglass shadow of the falling axle covered Cole as he stumbled up the off-ramp. The gyrating mass of red-hot metal passed over his left shoulder, slapping the length of tire tread against his back, driving him to the pavement. His shattered left shoulder and fractured arm whipped around his torso before he felt the pain. Driven hard forward by the impact, he fell once again, returning his broken nose to the freeway pavement. In slow motion, Cole saw the intricate detail of the road rising to meet him. Colorful little stones encased in the pavement came so close to his face that they blurred until turning his world black.

CHAPTER 19

HOMECOMING

Consciousness for Cole Walker came from the morning light stabbing through the cracks of the window blinds of his hospital room. The bright light burned his eyes, forcing him to squeeze them shut. His back was stiff, his arm ached, his chest hurt, his face throbbed. His mind spun with a swirl of unfamiliar faces, masked and muffled, fading in and out from the narcotic's waning grasp. Slowly, he began to remember more familiar faces: Barker, Val, Raz and Putnam. He squeezed his eyes tight and tried to remember where he was—and why.

Anticipating the blinding light, Cole cracked his eyes to a squint. The first thing he saw was the fuzzy white tape across the bridge of his nose, blurring his vision. He patted the tape with his fingertips, but the intense pain made his eyes water. He left his nose alone, deciding instead to look around. The room was painted stark Navajo and on the far wall was a small, mounted television. There were bouquets of flowers and balloons on the nightstand, the other bed, the windowsill and floor. A large teddy bear, wearing a yellow turnout jacket sat in a wheelchair. On the nightstand sat an array of *Get Well* cards surrounding a plate of cookies topped with a striped bow. A yellow note was teepee'd on the mound of cookies, *SAFE TO EAT – EVELYNN.*

Cole eased back into his pillow. Resting across his chest was his left arm in a bright white cast from shoulder to wrist. The cast lay heavy and uncomfortable across his chest. He tried to move it but stopped for the pain. With even slight movement, his body ached. He heard voices and footsteps passing in the hall. Turning to listen, he wracked his body with stabbing pain, giving him an instant headache.

At first, his awakening was surreal until the recesses of his foggy memory brought back the incident with the fuel tanker and the explosion that had happened just the other... *the other what?* Cole couldn't remember how long ago the explosion was. He had no grasp of time. *How long has it been?* His eyes darted around the drab room, searching for a clock or calendar or some evidence of time, but he saw nothing. A soft creak caught his attention as the door to the room slowly opened. He turned quickly to look, wracking his body with more pain. He froze in place, scared to move. Coming into focus through his bloodshot peripheral vision was the freckled nose and wavy red hair of Red Hall.

"You're finally awake," said Red, shoving the door open and entering the dull room, brightening it with his smile. He was wearing a dark blue uniform, not the standard PIO uniform of white shirt and black slacks—but a field uniform. Red stood beside the bed and squeezed Cole's good hand. It hurt too.

"It's nice to see you back with the living," said Red, flashing a broad smile.

"How's my crew?" asked Cole. His voice was raspy, his lips dry and cracked.

"They're all fine," said Red, giving Cole's hand another painful squeeze. "They're just worried about you."

Cole relaxed and slowly returned his head to the pillow, wincing from the slightest movement of his arm. "Was anybody hurt? Bleve or the lady?"

"Not a soul," said Red. "And we caught the whole thing on the five o'clock news. You're a hero, especially that stunt with the dog. The animal people ate it up."

Cole grinned with relief, his eyes moistened. It was the closest he had ever come to...He shook the thought and changed the subject. "How long have I been here?"

"About three days."

Cole turned cautiously, not wanting to offend any of his injuries, and studied his friend. A pale ring of white skin outlined Red's hairline from a recent haircut and on his uniform collar the hospital's cold fluorescent lights sparkled bright off shiny silver buttons. His collars bore captain insignias. "You're wearing bugles," said Cole, surprised.

"The Old Man promoted me yesterday." Red's eyes beamed as he flipped his collars with his thumbs. "It was either because the Brookfield town council unanimously ratified a ten-year contract for Los Angeles Fire protection, or because Mayor Jenkins resigned. Either way, I start my new assignment today." He rocked back on his heels, allowing his words to sink into Cole's bruised ears. "Do you have any advice for a new captain?"

"Don't drink from black coffee mugs," smirked Cole, causing his face to hurt.

Red dismissed the comment, blaming it on the Demoral.

Cole painfully held out his hand with a grimace disguised as a smile. "Congratulations, Red. What's your assignment?"

Red cleared his throat before giving Cole an answer. "Station Forty-five," he said, carefully watching Cole's reaction.

"Which rig?"

"The Truck."

"What shift?"

"C shift."

"Truck Forty-five, C shift!" blurted Cole, compounding his confusion. "That's Decker's assignment. Where'd Decker go?"

Red's answer came slow. He stammered his words and cleared his throat several more times before stuttering his answer. "He...he was transferred to your spot at Station Sixty-one."

Cole paused for a moment to visualize what Red had said. Cole's expression eroded to anger.

"Come on," said Cole, sitting up, grimacing as he jerked the bed sheets off his legs. "Get me out of here."

"Wait a minute," said Red, pressing firmly against Cole's good shoulder. "What are you planning to do? Kick his ass? Let's talk about this." Red stepped across the room to the medical chart box and retrieved the physician's clipboard. He put on his sunglasses, implying reading glasses, and flipped through Cole's medical chart.

"In case you haven't noticed, you're all screwed up. Ahh, here it is," said Red, stabbing his finger on the chart. "You have a comminuted humerus and scapula fractured into a gazillion pieces, requiring two pins and a roll of duct tape to hold them together." Red dipped his head and glanced at Cole over the rim of his sunglasses, then returned his attention to the medical chart. "Oh, look, there's more. You have three cracked ribs, a broken nose, spinal bruising and you've been unconscious for the past three days. If you want a second opinion, I'll add that you're hideous to look at and need an enema."

"Are you going to help me or am I going by myself?" asked Cole, as he pulled the IV from his arm. He swung his legs to the edge of the bed, jolting his body with pain, but controlled his expression not wanting Red to notice his discomfort.

Red shook his head, realizing that he'd lost the argument. He pushed the teddy bear out of the wheelchair and helped Cole sit down. On their way out of the room, Red spun the wheelchair abruptly facing Cole directly into the wardrobe mirror. Cole was stunned at what looked back. The tape that he couldn't get out of his vision was strapped across a purple and yellow bulbous nose. Both of his eyes were set in dark purple cavities, the outer edges were laced with yellow ecchymosis. His left cheek and forehead were scabbed. Purple fingertips protruded from the end of the huge cast, lying weighty on his lap.

"Let's go," said Cole, turning from the mirror.

They snuck out the side door of the hospital, not wanting to pass the nurses' station. Red wheeled Cole into the service elevator, then to his red sedan parked in the fire lane. Together, they struggled to put Cole in the front seat. The back seat was full of bedding and books and uniforms for Red's move to Station 45.

While Red drove toward Brookfield, Cole, using his one good hand, wrestled and wiggled into a pair of Red's uniform pants. He didn't bother with a shirt. He just synched the hospital gown around his shoulders. Every bump and pothole sent stabbing pains through Cole's shattered arm, which was not his biggest concern. Beyond the pain, all he could think about was Captain Decker in *his* firehouse.

They slowed to a stop in front of the handsome red brick fire station. The front bay doors were closed. Cole had Red pull around to the rear. From the back parking lot, he could see inside the apparatus room. The crew was standing at attention, lined up between the paramedic truck and the new Kovatch fire engine. Captain Decker was pacing back and forth, lecturing—just as he had lectured in the academy. The sight gave both Cole and Red a reminiscent shiver.

The back apparatus door was open, but the overpowering ramble of Captain Decker must have drowned out any noise from the sedan. Cole painfully exited the car, thankful that nobody was watching. Once standing, he straightened his posture and brushed his hand through his matted hair, then slammed the door. The noise of the car door broke the crew's trance on Captain Decker. They immediately broke file and headed for Cole, leaving Captain Decker mid-sentence. Leading the pack in full sprint, Putnam headed directly for the fragile and terrified Captain Walker. Putnam bore down on Cole, spreading his arms like he was tackling a quarterback. Cole stopped him with a straight arm and offered a handshake instead. The big man shook Cole's hand so vigorously that his other arm rattled inside the cast and hurt like hell, but Cole said nothing to stop him. The rest of the firemen took their turns with gentler handshakes.

For the moment, gathered in their semicircle around Captain Walker, they were a crew again. They talked and laughed and reminisced about their close call. Bleve circled the group, and with each burst of laughter, he barked in participation. So much had happened in the past three days.

"Did you hear Brookfield extended the contract?"

"Did you hear Mayor Jenkins quit?"

"Did you see our new fire engine?"

Putnam pulled two taped *CATS* tickets from his shirt pocket. "Evelynn asked me on a date," he beamed. He hooked his thick arm around Valdez's neck and scrubbed his thick pompadour with his knuckles. "Somebody was playing match-maker and gave her these tickets."

"Well, Val. Now that all LA Fire stations are wide open to you, what's your plan?" asked Cole.

"Well, Cap. I think I'm going to hang around and help Terry through paramedic school and maybe give him a few pointers on women. It's a big job, but somebody's got to do it."

"How about you, Raz?"

"I'll do the same. Then, I'll transfer to a station in Malibu next summer and maybe cross-train as a lifeguard."

Rus Barker gently shook Cole's hand again, obviously aware of his discomfort. Rus looked different, somehow brighter. Cole noticed his head was a little shinier. Long strands of hair no longer bridged his bald spot. He had a new haircut. Barker rubbed his smooth head, "I'm not trying to hide it anymore."

Cole nodded toward the new Kovatch fire engine, "Sorry about your fire engine."

"That's okay, Cap. That's how she would have wanted to go," said Barker, smiling.

"Let's see," said Cole looking up, rubbing his chin, showing some thought. "You only have six months to learn how to drive that thing."

"First of all, that *thing* is a s*he* and I have five months and 27 days to learn to drive *her* before I retire," said Barker. Both men laughed. Cole's laughter brought tears to his eyes for a variety of reasons. Barker said if Cole could get his arm mended within the next five months and 27 days, he'd welcome the chance to work together again. If not, he'd settle for a barbecue. It was the best compliment Cole could have received. He accepted.

Captain Decker wedged his way through the ring of firemen. The crew stepped aside, eventually migrating behind their new captain. Still hard as brass, Captain Decker studied Cole, sipping pensively from his black *Captain* coffee mug. Captain Decker offered a salutation, or greeting, or something, but Cole didn't pay much attention. His eyes were glued to Decker's blackened lips. The firemen mimicked drinking coffee behind Decker's back. Cole curbed his laughter by staring down at his bare feet, which Captain Decker interpreted as a sign of respect.

"Well, Captain Walker, I'm glad you made it out alive." Captain Decker turned to the crew, almost catching them in mid-gesture. "Social time is over, ladies. Make yourselves scarce." The firemen dispersed and busied themselves around the equipment but watched the two captains from a safe distance.

"They're a good crew," said Cole, struggling to contain his laughter, stealing glances of the large ebon smudge across Captain Decker's face.

Ding…

"Sure, they are," said Captain Decker, not convinced, glowering at the alarm bell.

Ddddrrrriiiinnnngggg!

The dispatcher's voice boomed through the apparatus room, announcing a fire. The firemen scrambled to their respective positions and donned their turnouts. Captain Decker sashayed to the fire engine. Cole followed. Decker stepped into his rubber boots, pulled up his turnout pants and blindly buckled the waist snap as he held his penetrating stare at Cole. Decker pulled his suspenders up and released them against his shoulders with a decisive snap. He shouted at the firemen, "Come on, ladies! Same day service!"

Cole clenched his good fist in contempt until he noticed Captain Decker standing tall in a pair of familiar pink panties. Captain Decker, obnoxious and arrogant, turned to Cole, "Don't worry, Walker. I'll have your crew in shape in no time."

Captain Decker climbed into the cab of the fire engine, flashing his *LA's FINEST* in large block letters across his butt, then slammed the door. Cole knocked on the captain's door. Decker rolled down the window, leaned out slightly, and looked down on Cole.

"Of all the possible captains, I'm glad you got this assignment," smiled Cole, raising the thumb of his good hand. Decker returned Cole's smile with a pompous, egotistical, black-lipped smirk, then pointed forward signaling Barker to drive.

Engine 61 roared out of the fire station, followed by the paramedic truck. Cole watched their flashing red lights until they disappeared from his blurry vision. He swallowed hard against the knot in his throat. Captain Cole Walker stood alone in the empty firehouse with his eyes squeezed tight and allowed the fading sirens to burn into his memory.

About the Author

Michael R. Jasperson has worked over 35 years in the fire service. He has served as a firefighter, paramedic, engineer, inspector, Haz Mat specialist and fire captain. Once co-editor of *Frontline Magazine*, he now writes a column for *Straight Streams,* a monthly fire service magazine. Questions and comments can be directed to: thoughttorpedo.com

OTHER TITLES BY MICHAEL R JASPERSON

What My Parents Couldn't Teach Me,
I had to Learn from My Children